BG Chemie

Toxicological Evaluations

6 Potential Health Hazards of Existing Chemicals

Springer-Verlag
Berlin Heidelberg New York
London Paris Tokyo
Hong Kong Barcelona
Budapest

Berufsgenossenschaft
der chemischen Industrie
Kurfürsten Anlage 62

69115 Heidelberg
FRG

ISBN-13:978-3-642-84994-7 e-ISBN-13:978-3-642-84992-3
DOI: 10.1007/978-3-642-84992-3

Library of Congress Cataloging-in-Publication Data. (Revised for volume 6)
Toxicological evaluations. Includes bibliographical references and index.
1. Toxicity testing. I. Berufsgenossenschaft der Chemischen Industrie.
[DNLM: 1. Hazardous Substances toxicity. WA 465 T7549] RA1199.T678
1990 615. 9'07 90-10020
ISBN-13:978-3-642-84994-7

Softcover reprint of the hardcover 1st edition 1993

51/3130-5 4 3 2 1 0 – Printed on acid-free paper

Preface

As part of its "Programme for the prevention of health hazards caused by industrial substances", the Berufsgenossenschaft der chemischen Industrie (BG Chemie, Employment Accident Insurance Fund of the Chemical Industry) began in 1977 to investigate the toxicity of those chemicals which are widely used, have many different applications and are suspected of being dangerous to health, in particular of having long-term effects. The investigations consist of a literature search and – depending on the results – commissions of experimental studies. It is hoped by means of this testing to close gaps in our knowledge and to increase the scientific validity of the required risk assessments. The results of the toxicological investigations carried out by BG Chemie, and the resulting substance assessments have been published in German since 1987 in the form of 132 „Toxikologische Bewertungen" ("Toxicological Evaluations") up to now.

In order to make this useful information internationally available, BG Chemie began in October 1990 to publish them as a book series in English, of which the sixth volume (containing 11 individual evaluations) is presented here. Therefore for 83 existing chemicals "Toxicological Evaluations" are available in English at the moment, a further 27 are in preparation and will be published soon.

Because of the short time between publishing volume 1–6, printing of "Introduction" (consisting of a general overview of the programme, lists with names of people involved as well as substances under investigation) was abandoned in volumes 2, 3, 5, and 6. If more detailed information is required about the ongoing work, see volume 1 or 4 or contact BG Chemie at first hand.

BG Chemie hopes that, for many people working in the chemical industry, this information will be of practical help in assessing hazards to health at the workplace.

Contents

Monochloroacetic acid

1. Summary and assessment

After absorption, monochloroacetic acid is rapidly distributed in the body. It is excreted unchanged in a fast elimination phase (rat, subcutaneous, half-life 90 minutes) and is bound to glutathione, cysteine or protein or is eliminated as CO_2 in a slow phase (half-life 500 minutes). In rats, about 50% of the radioactivity is excreted in the urine after 17 hours, while the corresponding figure in mice after 72 hours is about 82–88%. In humans, after contamination of the skin with ^{14}C-labelled monochloroacetic acid, a half-life of about 15 hours has been found for the rapid excretion of monochloroacetic acid in the urine, but no information has been provided on the slow phase of elimination. After 6 days only a small amount is detectable in the blood. In vivo studies on mice have revealed that monochloroacetic acid is metabolized to thiodiacetic acid (via S-carboxymethyl glutathione and S-carboxymethyl cysteine) and, to a lesser extent, to glycolic acid and CO_2 (after enzymatic hydrolysis of the C-Cl bond).

Monochloroacetic acid is acute toxic (LD_{50} rat oral ca. 90 mg/kg). After oral exposure, systemic effects in laboratory animals and in domestic animals (cows, sheep, poultry) include weight loss, tremors, convulsions, paralysis of the extremities, histologically detectable changes in the central nervous system and bleeding in the lungs and heart muscle. Monochloroacetic acid is easily absorbed through the skin in toxic amounts. The vapours evolved at 20 °C appear to have no obvious toxic effects, but the aerosols formed at higher temperatures can be fatal. The repeated application (gavage) of 120 mg/kg for 14 days is mainly tolerated by rats and mice. Doses of 240 and 480 mg/kg, however, are lethal for mice.

Monochloroacetic acid has corrosive effects on the skin and eyes.

There are no indications that monochloroacetic acid has skin sensitization potential.

Subchronic administration of monochloroacetic acid to rats and mice by stomach tube at doses of up to 200 mg/kg, or to rats in their drinking water, led to changes in clinical-chemical parameters in rats which are indicative of nephro- and liver toxicity. Histologically, effects are observed in the liver, kidneys and lungs. In mice on the other hand, no treatment-related effects on haematological or clinical-chemical parameters are observed. Administration in the feed for 30 weeks at a concentration of 100 mg/kg does not cause any pathological effects in rats, except for a reduction in weight. In an inadequately-documented 4-month inhalation study, rats and guinea-pigs exposed at 20.8 mg/m^3 show beside other effects reduced weight gain, a fall in haemoglobin levels, haemoglobinaemia and histologically-detectable inflammatory changes in the respiratory tract. A concentration of 5.8 mg/m^3 has no detectable effect.

There is no indication of genotoxic potential in bacterial mutagenicity studies with and without S9-mix. In a test of the mutagenic effect in mammalian cells in vitro, the results are contradictory. The results of a Drosophila-test are equivocal. After intraperitoneal application an increase of chromosomal aberrations is observed in mice.

In long-term studies in various mouse strains treated dermally, subcutaneously or orally, there are no indications that monochloroacetic acid is carcinogenic. An oral (gavage) 2-year study conducted within the scope of the NTP in rats and mice shows no evidence of carcinogenicity. The finding that monochloroacetic acid does not bind covalently to cellular structures support also the arguments against a carcinogenic effect.

After administration of monochloroacetic acid to pregnant rats at doses of up to 140 mg/kg, the numbers and weights of foetuses are comparable with control values. Malformations of the cardiovascular system are significantly increased at the highest dose level, but skeletal malformations are not observed. The intraperitoneal application of up to 50 mg/kg causes an increased number of abnormal sperms in male mice.

In vitro studies show that monochloroacetic acid inhibits acetate oxidation and gluconeogenesis. Liver glutathione-S-transferase is dose-dependently inhibited by direct binding of monochloroacetic acid to the enzyme.

On human skin monochloroacetic acid causes irritation, burning (with blister formation) and corrosion. Similar damage is to be expected on contact with the eye. Monochloroacetic acid is rapidly absorbed through the skin and, if a sufficiently large area of the skin is contaminated, can lead to breathing difficulties, convulsions, vomiting, shock and death.

2. Name of substance

2.1	Usual name	Monochloroacetic acid
2.2	IUPAC-name	Monochloroacetic acid
2.3	CAS-No.	79-11-8

3. Synonyms, common and trade names

Chloroacetic acid
Chloroethanoic acid
MCA
MCAA
Monochloressigsäure
Chlorethansäure

4. Structural and molecular formulae

4.1	Structural formula	$Cl\text{-}CH_2\text{-}COOH$
4.2	Molecular formula	$C_2H_3O_2Cl$

5. Physical and chemical properties

5.1	Molecular mass, g/mol	94.50
5.2	Melting point, °C	Commercial α-form: 62.3-62.8 β-form (unstable): 56.5 γ-form (unstable): 50.6 (Koenig, 1975)
5.3	Boiling point, °C	189.4 (Koenig, 1975)
5.4	Vapour pressure, hPa	0.2 (at 20 °C) 2.0 (at 50 °C) (Hoechst, 1991)
5.5	Density, g/cm³	1.3703 (at 65 °C) (Koenig, 1975)
5.6	Solubility in water	86.0 g/100 g solution (at 30 °C) (Koenig, 1975)

5.7 Solubility in organic solvents	Dissolves well in methanol, ethanol, acetone, ether, dioxane, dimethylformamide, dimethyl sulphoxide; slightly soluble in hydrocarbons and chlorinated hydrocarbons (Koenig, 1975)
5.8 Solubility in fat	No information available
5.9 pH-value	About 1 (10 g/l water) (Koenig, 1975)
5.10 Conversion factor	1 ppm ≙ 3.92 mg/m³ 1 mg/m³ ≙ 0.26 ppm (at 25 °C and 1013 hPa)

6. Uses

Used in the production of carboxymethylcellulose. Intermediate in the manufacture of pesticides and other products (Hoechst, 1991).

7. Experimental results

7.1 Toxicokinetics and metabolism

In rats that were given 1 μCi (37 kBq) 1-^{14}C-monochloroacetic acid orally, the radioactivity in the plasma, liver, kidneys, heart, testes and spleen reached a maximum after 1 to 2 hours and thereafter declined rapidly (half-life 2 to 7 hours). The radioactivity in the brain however, although present at lower levels than in the other organs, increased for up to 8 hours and then remained at a plateau for 24 hours (Berardi and Snyder, 1983).

In two distribution studies with ^{14}C-labelled monochloroacetic acid in Sprague-Dawley rats, higher levels of radioactivity were measured in the liver and kidneys than in the plasma 2 hours after subcutaneous injection of the LD_{90} (24 hours, 162 mg/kg), while the levels in the heart and the brain were comparable with plasma levels. Subcutaneous injection of 53 mg/kg (=LD_1) led to a similar distribution. The maximum plasma concentration was detected after

32 minutes. From the 2-phase elimination curve, the plasma half-life was determined to be 90 minutes for the fast phase and 500 minutes for the slow phase. After 17 hours, approximately 50% of the administered radioactivity had been excreted in the urine (Hayes et al., 1972, 1973).

Female albino mice received intraperitoneal injections of ^{14}C-labelled monochloroacetic acid with an activity of 2 to 4 mCi at doses of 0.07, 0.09 and 0.1 g/kg. The amount of ^{14}C in the urine, expired air and faeces was determined after 24, 48 and 72 hours. Within 72 hours, 82–88% was found in the urine, 8% in the expired air and 0.2–0.3% in the faeces (which were contaminated with urine). The amount remaining in the body was 2–3%. Two major metabolites found in the urine by paper chromatography were S-carboxymethyl-L-cysteine (33–43% as the free compound and 1–6% conjugated) and thiodiacetic acid (33–42%). Also detected were monochloroacetic acid at 6–22%, glycolic acid at 3–5% and oxalic acid at 0.1–0.2%. These quantities were estimated by isotopic dilution analysis (percentage activity in the urine). The authors assumed there were two different pathways for the breakdown of monochloroacetic acid in the organism: the main route was the formation of S-carboxymethyl glutathione and subsequently S-carboxymethyl cysteine, which would then be metabolized to thiodiacetic acid. A second, less important, metabolic pathway was enzymatic hydrolysis of the C–Cl bond, with subsequent formation of glycolic acid which would be mainly oxidized to CO_2 (Yllner, 1971).

Male Sprague-Dawley rats were injected in the tail vein with [1-^{14}C]-monochloroacetic acid at a dose of 6.8 µg/100 g body weight, which was equivalent to an activity of 40 µCi (1.48×10^6 Bq). After 5 minutes and after 1, 4, 12, 24 and 48 hours, 3 rats at each time interval were embedded in carboxymethylcellulose and then frozen. Whole-body autoradiographs were then prepared from longitudinal sections of the frozen animals. Analysis of the developed sections showed that at 5 minutes there was a rapid accumulation of radioactivity in the liver, kidney cortex, stomach wall, salivary and tear glands, oesophagus, tracheal tissues, pancreas and certain ganglia of the peripheral nervous system, and some accumulation in the brain. The radio-activity disappeared rapidly from the blood circulation. One hour after the administration of ^{14}C-labelled monochloroacetic acid, the radioactivity was extensively eliminated into the small intestine, the kidney contents and the urinary bladder. A high degree of accumulation was observed in the brain, spinal cord, thymus gland, myocardial tissue, salivary glands and tongue. After 4 hours, the liver and various other tissues began to eliminate the radioactivity, while the brain (in particular the cerebellum), the spinal cord, the thymus and the pancreas had accumulated more monochloroacetic acid, and still continued to do so after 48 hours. These results support an early accumulation of monochloroacetic acid and/or its metabolites in hydrophilic tissues, and subsequent accumulation in lipophilic tissues (Bhat et al., 1990).

In a laboratory accident, a worker's finger was contaminated with ^{14}C-labelled monochloroacetic acid (about 60–70 mc, estimated dose about 200–1600 rad). He washed his hands within one minute. The degree of skin contamination and the concentration of ^{14}C in the blood, urine and expired air was measured. The β-radiation measured on the skin (β-emissions/cm^2/second) fell from 10 000 immediately after contamination to 0, within 10 days. A rapidly-excreted substance was found in the urine (half-life ca. 15 hours), and this appeared to be mainly unchanged monochloroacetic acid. In the slow phase (no information on half-life) monochloroacetic acid which had reacted with cysteine, glutathione or protein was excreted in the urine (ca. 330 µc, equivalent to approximately 0.002 ml labelled monochloroacetic acid). A similar amount of radioactivity was eliminated as CO_2. Less than 20% of the radioactivity was found in the erythrocytes and about 80% in the plasma, 17.5 hours after the accident. After 6 days, only small amounts (0.16 µc/ml) could be detected in the blood (Dancer et al., 1965).

7.2 Acute and subacute toxicity

LD_{50} values for monochloroacetic acid from various exposure routes are given in Table 1 (page 8, 9).

After administration of the substance, apathy and weight loss were observed (Woodard et al., 1941), and in a further study motor disturbances, loss of balance and lying on the stomach were also seen (Hoechst, 1979 a). Similar signs of toxicity were observed on dermal application to rats and rabbits, in addition to local tissue damage (Hoechst, 1979 b, c). After subcutaneous administration of monochloroacetic acid, rats suffered tonic-clonic spasms, polydipsia and depressed breathing. Pathological examination revealed swelling of the liver and spleen (Hayes et al., 1973) and necrosis at the site of injection (Hoechst, 1979 d).

When monochloroacetic acid in neutral solution was given by stomach tube to male and female Sprague-Dawley rats as a single dose of 94 mg/kg, all animals survived the 3-day observation period. There was no change in body weight, urine volume or urine composition (osmolality, sodium, potassium, glucose, protein) compared with the controls. At 282 and 470 mg/kg, 1/3 of the male rats survived for 24 hours. In the animals that survived the 470 mg/kg dose, urine volume and osmolality were unchanged but the sodium concentration was increased. Potassium was increased in the plasma, and liver glutathione was reduced. Similar effects were not seen on administration of 282 mg/kg (Davis and Berndt, 1987).

LD_{50} values of 300 mg/kg after oral administration and 150 mg/kg after subcutaneous application have been reported for male mice. Signs of toxicity seen in the mice prior to death included tremors, tonic-clonic spasms and respiratory paralysis. A few of the surviving mice exhibited the Straub tail effect, severe tremors and, after 24 hours, paralysis of the forelegs (Berardi and Snyder, 1983).

Calves and sheep died after drinking water which had been accidentally contaminated with monochloroacetic acid. The minimum lethal dose for calves and sheep was estimated to be about 20–70 mg/kg, but is more likely to have been around 170 mg/kg (estimated from the amount of water drunk and the concentration of monochloroacetic acid in the water). Both species showed paralysis of the extremities, tremors and convulsions. Autopsy examination revealed congestion of the lungs, subcutaneous bleeding and oedema, and haemorrhage and bleeding in the endocardium and epicardium (Quick et al., 1983).

In a controlled study, cows were lethargic for 24 hours after oral administration of 50 mg sodium monochloroacetate/kg. Twice the dose (100 mg/kg) caused marked signs of toxicity (loss of appetite, diarrhoea, muscle tremors and loss of tone in the rumen musculature). The animals took 2 weeks to recover. Death occurred 9 hours after administration of 150 mg/kg. Oral doses of 5 to 50 mg sodium monochloroacetate/kg for 28 days were without effect in cows (Dalgaard-Mikkelsen and Rasmussen, 1961).

Vomiting and increased bile secretion were observed in dogs after oral administration of 24 mg/kg. The same symptoms occurred after intravenous administration of 25–30 mg/kg (Emerson and Morrison, 1943; Dalgaard-Mikkelsen et al., 1955).

Table 1. LD_{50} values for monochloroacetic acid

Species	Sex	Route of administration	Observation period	LD_{50} (mg/kg)	References
Rat	–	Oral, as Na salt	–	580[a]	Maksimov and Dubinina, 1974
Rat	–	Oral	–	277.5	Kurĉatov and Vasileva, 1976
Rat	–	Oral, as Na salt	6 days	76[a]	Woodard et al., 1941
Rat	♀	Oral, 1% in water	14 days	90.4 (73.6–112)	Hoechst, 1979a
Rat	–	Oral, as Na salt	–	410[a]	Babanov et al., 1984
Mouse	♂, ♀	Oral, as Na salt	6 days	255[a]	Woodard et al., 1941
Mouse	–	Oral, 5% solution as Na salt	5 days	165[a]	Morrison, 1946
Mouse	–	Oral, as Na salt	–	415[a]	Babanov et al., 1984
Mouse	♂	Oral	–	300	Berardi and Snyder, 1983
Mouse (Swiss-Webster)	–	Oral	–	260	Berardi et al., 1987
Guinea-pig	♂	Oral, as Na salt	6 days	79[a]	Woodard et al., 1941
Rat	♀	Dermal, as 40% dilution in 0.9% NaCl	14 days	305 (242–384)	Hoechst, 1979b

Rabbit	–	Dermal, as 50% solution in 0.9% NaCl	14 days	ca. 250	Hoechst, 1979c
Rat	♀	Subcutaneous, as a 50% solution in 0.9% NaCl	–	97.4 (89.9–105.5)	Hoechst, 1979d
Rat (Sprague-Dawley)	♂	Subcutaneous	24 hours	108	Hayes et al., 1972, 1973
Mouse	♂	Subcutaneous	–	150	Berardi and Snyder, 1983
Rat	–	i.v., 10% solution	–	55	Maksimov and Dubinina, 1974

– Strain and/or observation period not specified

[a] Administered as the sodium salt, calculated as the acid

2 geese tolerated the administration of 50 mg sodium monochloroacetate/kg by stomach tube. These animals were then given 75 mg/kg 1 week later, and both died within 4–6 hours. After 3 hours, incoordination and convulsions occurred. Post-mortem examination revealed redness and point-like haemorrhages in the digestive tract (Christiansen and Dalgaard-Mikkelsen, 1961).

Deaths occurred following dermal application to rats of 200 or 400 mg monochloroacetic acid/kg for 10, 20 or 40 minutes (Berardi and Snyder, 1983).

A dermal LD_{50} of 250 mg/kg was reported in rabbits for monochloroacetic acid (Hoechst, 1979 c).

Deaths occurred in rabbits when 3% of the body surface area was exposed to monochloroacetic acid; when the skin was washed 1 minute after exposure no decrease in the mortality rate was found (no further details; Christofano et al., 1970).

Male white New Zealand rabbits (weight 2 to 2.5 kg, 5 to 10 animals/group) had molten monochloroacetic acid (ca. 60 °C) applied to their clipped dorsal skin (3 to 5% of the surface area of the body) for 1.5 or 5 minutes. This was then washed off under warm tap water for 2 minutes. Immediately after the substance was wased off, a proportion of the animals had ethanol (0.5 to 3 g/l in physiological saline) infused intravenously into the ear. The application of monochloroacetic acid caused the death of all animals within a few hours. Signs of toxicity observed included loss of balance, lethargy, progressive shortness of breath and coma. As with monofluoroacetic acid intoxication, hyperglycaemia and severe acidosis were observed. Treatment of the rabbits with ethanol did not affect the mortality in this study, but death occurred later in the treated animals. The blood sugar level was also affected less in the treated rabbits. The authors considered that, as with monofluoroacetic acid intoxication, cis-aconitase in the citric acid cycle was inhibited by chloroacetate, causing the accumulation of citric acid and consequent metabolic acidosis. Ethanol, occurring concurrently with chloroacetate, would be expected to act as an acetate donor (Millischer et al., 1987a).

When the tails of mice were dipped in a 10% solution of monochloroacetic acid (solvent not specificed), toxic effects occurred in 50% of the animals after 35 minutes (no further details; Maksimov and Dubinina, 1947).

Dynamic inhalation of the maximum concentration of chloroacetic acid vapour which could be generated at 20 °C (5 mg/m^3) did not cause deaths in rats. To obtain higher concentrations, the chloroacetic acid solution was heated to 95 °C. With the condensed monochloroacetic acid aerosol an LC_{50} value of 180 (146–221) mg/m^3 was obtained, most of the deaths occurring within the first 24 hours. The same authors reported that 17.3 mg/m^3 caused no changes in rectal temperature, oxygen consumption, lung colour or breathing rate in rats. Increased breathing frequency was observed from 23.7 mg/m^3 (threshold concentration). From 93.6 mg/m^3 acute damage occurred (no details). No information was provided on the duration of exposure for any of the experiments (Maksimov and Dubinina, 1974).

A threshold concentration of 24.7 mg/m^3 was reported for irritant effects on the respiratory tract of rats (no data on duration of exposure; Rodionova and Ivanov, 1979).

Inhalation of monochloroacetic acid vapour for 1 minute (concentration not specified) did not cause any systemic toxicity in rats (Christofano et al., 1970).

In a 14-day study, male and female F344/N rats (5 animals/sex/dose) were given monochloroacetic acid in deionized water at doses of 0, 7.5, 15, 30, 60 or 120 mg/kg body weight by stomach tube a total of 12 times. 1 male rat at the highest dose level died during the study. Clear nasal secretions and/or watering of the eyes were observed in all groups treated with monochloroacetic acid. No treatment-related macroscopically detectable lesions were observed (NTP, 1990).

In an analogous study in B6C3F1 mice, the males were given monochloroacetic acid at doses of 0, 15, 30, 60, 120 or 240 mg/kg, and the females were given doses of 0, 30, 60, 120, 240 or 480 mg/kg body weight. All of the males that received 240 mg/kg and all of the females that received 240 or 480 mg/kg died during the study. In the animals in the highest dose groups (240 mg/kg for males, 480 mg/kg for females) ruffled fur, hypoactivity, ataxia and watering of the eyes were observed. No treatment-related macroscopic lesions were seen (NTP, 1990).

7.3 Skin and mucous membrane effects

Concentrated monochloroacetic acid solutions (no information on solvent, concentration, exposure duration or protocol) caused severe chemical burns to rabbit skin. The threshold concentration for rabbit skin was reported to be 0.05% (Maksimov and Dubinina, 1974).

A 10% aqueous solution of monochloroacetic acid caused redness and oedema when applied semi-occlusively to the skin of 18 Chinchilla rabbits. A 1% solution applied under the same conditions caused no changes to the skin (Rodionova and Ivanov, 1979).

Sodium monochloroacetate was said to have no particular irritant effect on the skin of rats on single or repeated application (Babanov et al., 1984).

Covered application of 500 mg monochloroacetic acid (made into a paste with 0.05 ml 0.9% NaCl solution) to the intact or abraded clipped skin of white Himalaya rabbits (2 application sites) with a 2.5 cm^2 patch (no precise details) was fatal to all of the rabbits. A 24-hour application to the intact skin of 100 mg/kg as a 50% solution (with a 2.5 cm^2 patch) was tolerated, but led to very marked erythema and oedema, as well as grey-green necroses which sloughed off within 48 to 72 hours (Hoechst, 1979 e).

Concentrated monochloroacetic acid solutions (no precise details) caused severe chemical burns in rabbits' eyes (Maksimov and Dubinina, 1974).

The instillation of 100 mg monochloroacetic acid (made into a paste with 0.01 ml physiological saline) into rabbits' eyes caused very severe irritation and grey burns of the conjunctiva (Hoechst, 1979 e).

7.4 Sensitization

An inadequately documented study reported no evidence of sensitization potential in rabbits when one drop of a 0.1, 1.0, 5.0, 10.0 or 50.0% solution was applied to the skin (solvent and duration of treatment not specified), followed by a challenge 30 days later with 10 drops of 5% solution applied to the same site (Maksimov and Dubinina, 1974).

7.5 Subchronic and chronic toxicity

Groups of 20 male and 20 female F344/N rats were given an aqueous solution of monochloroacetic acid daily by stomach tube at doses of 0, 30, 60, 90, 120 or 150 mg/kg body weight, 5 times a week for 13 weeks. The following animals suffered treatment-related deaths at the three higher dose levels: all of the males given 120 and 150 mg/kg, 9 out of 10 males given 90 mg/kg, and all of the females given 90 to 150 mg/kg. The body weight gain in the surviving rats equalled that of the controls. There were dose-dependent increases in blood urea-nitrogen, in alanine aminotransferase and aspartate aminotransferase activities, and in the relative liver and kidney weights. The rats also showed a dose-related increase in the incidence and severity of cardiomyopathy, which the authors considered to be due to inhibition of mitochondrial aconitase activity in the heart. The no observable effect level for rats was given as 30 mg/kg body weight (Bryant et al., 1992).

In a study analogous to that described above, groups of 20 male and 20 female B6C3F1 mice were given doses of 0, 25, 50, 100, 150 and 200 mg/kg body weight. All of the males and 2 of the females in the highest dose group died. The body weight gain of the surviving animals was similar to that of the controls, with the exception of the females in the 200 mg/kg group, whose weight was significantly below that of the controls. No treatment-related effects on haematological or clinical-chemical parameters were observed in the mice. The absolute and relative liver weights were significantly increased in the females at the highest dose level. Mice in the high dose group that died during the study showed hepatocellular vacuolization. Based on these results, a no observable effect level of 100 mg/kg body weight was established for mice (Bryant et al., 1992).

Groups of 10 male and 10 female Sprague-Dawley rats were given the sodium salt of monochloroacetic acid in distilled water at doses of 15, 30, 60 or 120 mg/kg body weight. The substance was administered by stomach tube on 90 consecutive days. The control animals were treated with distilled water. At the highest dose level, 30% of the females and 80% of the males died, the majority within the first 2 days of treatment. The target organs were predominantly the liver and the kidneys. In rats that died within the first 3 days, haemorrhages and congestion of the lungs were observed (possibly a post-mortem effect), while in animals that died later, the predominant effect was liver damage. Increased blood levels of creatinine, calcium and urea-nitrogen indicated nephrotoxicity, and increases in the serum levels of alanine aminotransferase and aspartate aminotransferase were indicative of liver toxicity. The relative weights of both organs were increased. Microscopic examination revealed chronic nephropathy and increased pigmentation of the spleen in males that were given 60 mg/kg. The predominant effect seen in the males was lipidosis, and in the highest dose group, necrosis was also seen in the liver. A no effect level could not be established for the sodium salt of monochloroacetic acid in this study (Daniel et al., 1991).

Groups of 6 young male Wistar rats were given 0.005, 0.01, 0.025, 0.05 or 0.1% monochloroacetic acid mixed with their food (about 5, 13, 25, 50, or 100 mg/kg), or were given control food, for 208 days. Treatment-related clinical symptoms were few, neither were there many macroscopically or histopathologically detectable changes to the major organs (heart, liver, spleen, adrenal glands, kidneys, lungs, stomach, gut, pancreas, thyroid gland, nose and testes). Only the body weight gain of the top-dose rats was clearly reduced. In a further experiment, where the same authors fed 14 Wistar rats a diet containing 0.1% monochloroacetic acid for 90 days, motor activity in the running cage was reduced and there was a slight (questionable) increase in the glycogen content of the liver. The oxygen consumption (Warburg method) in slices from the liver, brain, kidney cortex and skeletal muscles was not increased (Fuhrmann et al., 1955).

5 male Sprague-Dawley rats (weight 220 to 242 g) were given monochloroacetic acid (1.9 mM) in their drinking water for 90 days, while 5 control animals were given double distilled water. The daily intake of monochloroacetic acid was 1/4 of the LD_{50} value of 76 mg/kg quoted by the authors. At the end of the study the body weights of the treated rats did not differ from those of the controls, but the liver weights were, however, ca. 10% lower. No adverse effects were detectable macroscopically, but microscopically, slight histopathological changes were found in the livers and lungs of the treated rats. The portal triads in the liver had distended portal veins, which were surrounded by an increased number of bile ductules, areas of oedema and variable numbers of inflammatory cells. In 4 of the rats, collagen deposition was slightly increased in a few portal triads. In the lungs, foci of perivascular inflammation, which contained lymphocytes and macrophages, were seen on the periphery of small veins. In addition, there was an increase in the thickness of the venous walls around these foci (Bhat et al., 1991).

In an inhalation study, 25 rats and 18 guinea-pigs (in total) were exposed over at least 4 months at concentrations of 20.8±1 and 5.8±0.3 mg monochloroacetic acid/m^3 (details of experimental protocol, daily exposure duration etc. were not supplied). The following parameters were studied: body weight, rectal temperature, blood morphology, liver and kidney function, oxygen consumption after swimming, respiratory rate, oxygen uptake of the lungs, organ weight and histology (no further details). At various (unspecified) times during the study, the animals in the high exposure group (20.8 mg/m^3) showed loss of body weight (rats, guinea-pigs), reduced oxygen consumption and body temperature, haemoglobinaemia, reduced chloride levels in the urine and increased specific gravity of the urine. Inflammatory changes were detected histologically in the respiratory tract (trachea, bronchi, lungs). The animals exposed at 5.8 mg/m^3 showed only a temporary reduction in the chloride content of the urine, and the remaining parameters were unaffected. Histologically, no differences were found in the respiratory tract, compared with controls. A recommendation for a Russian MAK value of 1 mg/m^3 was made from the results of this work (Maksimov and Dubinina, 1974).

7.6 Genotoxicity

7.6.1 In vitro

Monochloroacetic acid showed no mutagenic effects in *Salmonella typhimurium* strains TA 1530 and TA 1537 either with or without metabolic activation (Bartsch et al., 1975, 1980; Malaveille et al., 1975; McCann et al., 1975).

In another Ames test in *Salmonella typhimurium* strains TA 1535, TA 1537, TA 98 and TA 100, with and without S9-mix (Aroclor 1254-induced microsomes), monochloroacetic acid was similarly found to be inactive (Mortelmans et al., 1986).

In an Ames test in strains TA 98, TA 100, TA 1535 and TA 1537 with and without S9-mix, monochloroacetic acid was not mutagenic when tested up to a concentration of 1000 µg/plate (Hoechst, 1979f).

In a further study in *Salmonella typhimurium* strain TA 1535 without metabolic activation, monochloroacetic acid was negative at concentrations of up to 1000 mmol/plate. From 500 mmol/plate it was toxic to the bacteria (Rannug et al., 1976).

An indirect study of DNA damage with 2 strains of *Escherichia coli* (WP2 and WP100), gave negative results both with and without S9-mix (Mamber et al., 1983).

In an in vitro study on gene mutation with Chinese hamster V79 cells (in which the induction of 8-azaguanine-resistant and ouabain-resistant mutations was tested), there were no indications of mutagenic activity for monochloroacetic acid without S9-mix (Huberman et al., 1975).

There was no evidence of mutagenic potential at low concentrations (no exact details) in a modified mouse lymphoma test with S9-mix (Amacher and Turner, 1980).

In a later publication by the same authors, higher concentrations (330 to 784 μg/ml) up to the toxic dose range induced dose-dependent mutagenicity (Amacher and Turner, 1982).

In the absence of S9-mix, dose-dependent mutagenic activity was seen in the mouse lymphoma assay from 200 μg/ml (McGregor et al., 1987).

A significant increase in sister chromatid exchanges (SCEs) occurred in Chinese hamster ovary (CHO) cells on exposure to monochloroacetic acid at concentrations of 160 000 μg/ml or more without S9-mix. In the presence of S9-mix, no increase was observed at the same concentrations (Galloway et al., 1987).

In CHL cells (Chinese hamster lung fibroblasts), monochloroacetic acid did not induce chromosomal aberrations either with or without S9-mix at concentrations of 0.06 to 0.5 (toxic) mg/ml. At concentrations of 0.06 to 0.25 mg/ml without S9-mix no SCEs were found (Sawada et al., 1987).

In an in vitro study in rat bone marrow cells, monochloroacetic acid showed no significant inhibitory effect on DNA synthesis (thymidine uptake; Gross et al., 1982).

The induction of DNA repair by monochloroacetic acid could not be detected in the umu test in *Salmonella typhimurium* TA 1535/pSK1002 by measurement of β-galactosidase activity (Nakamura et al., 1987; Ono et al., 1991).

Monochloroacetic acid was negative in the SOS chromotest in *Escherichia coli* (Szegedi, 1989).

7.6.2 In vivo

Monochloroacetic acid (analytical pure) dissolved in distilled water was given intraperitoneally to male and female mice (Swiss strain, 10–12 weeks old) in doses of 50, 25, 12.5, or (5 × 10 mg/kg) body weight. 6 to 120 hours after application a significant increase of chromosomal aberrations was observed in the bone marrow of the animals. No increase of chromosomal aberrations, however, was observed 24 hours after oral or subcutaneous administration of 50 mg/kg (Bhunya and Das, 1987).

Monochloroacetic acid was tested for its ability to induce recessive lethal mutations in *Drosophila melanogaster*. When monochloroacetic acid (400 ppm in water) was administered in the feed for 3 days, no mutagenic effect was found. Application of monochloroacetic acid by injection (900 ppm in water) caused a slight increase in the mutation frequency. These results were, however, considered to be questionable (University of Wisconsin-Madison, 1990).

7.7 Carcinogenicity

In a screening test in which inhibition of interferon-induction in mouse embryo fibroblasts (after stimulation with Newcastle disease virus) serves as an indicator of carcinogenic activity, monochloroacetic acid was without effect (Sonnenfeld et al., 1980).

In a long-term study, groups of 50 female Swiss mice were treated dermally (3 times a week) or subcutaneously (once a week) for 580 days. Each dermal application was 2 mg monochloroacetic acid in 0.1 ml acetone/animal, and each subcutaneous dose was 0.5 mg in 0.05 ml tricaprylin. After dermal application, neither papillomas nor sarcomas occurred. In the group treated subcutaneously, the three sarcomas seen around the injection sites did not represent any significant increase ($p>0.05$). The average survival time for both treated groups was lower than that of the controls (dermal treatment, survival time 506 days, control 543 days; subcutaneous treatment, survival time 454 days, control 493 days; van Duuren et al., 1974).

18 males and 18 females each from two different mouse strains (strain X, a cross between C57BL/6 and C3H-Anf mice; strain Y, a cross between C57BL/6 and AKR mice) received 46.4 mg/kg as the maximum tolerated dose by stomach tube from day 7 to day 28, and then 149 ppm in their feed for 18 months. All of the animals were dissected, and the major organs of the thoracic and abdominal cavities were histologically examined, as were any lesions that were observed macroscopically. Blood smears were also taken before the animals were killed, but these were only evaluated in cases of splenomegaly or lymphadenopathy. In this study, monochloroacetic acid did not cause any significant increase in tumor incidence compared with controls (Innes et al., 1969).

Mice of the same strains (18 of each sex) received a single subcutaneous injection of 100 mg into the skin of the neck at the age of 28 days. There were no indications of a carcinogenic effect at the end of the study (no further details; Bionetics Research Labs, 1968).

In a 2-year study, monochloroacetic acid in deionized water was given to groups of 70 F344/N rats per sex at doses of 0, 15 or 30 mg/kg body weight and to groups of 60 B6C3F1 mice per sex at doses of 0, 50 or 100 mg/kg. In comparison with the controls, the high-dose male and female rats and the high-dose male mice showed significant reductions in survival time. The body weights of the male and female rats were up to 10% below those of the controls. No treatment-related increase in the incidence of neoplasms was observed in the rats. The increased occurrence of stroma polyps of the uterus in treated female rats compared with controls was within the values for historical controls. In the treated male mice, the average body weight was similar to that of the controls, while for the treated females it was 6 to 10% below control values. No increase in the occurrence of neoplasms was observed in the male or female mice. The increased occurrence of inflammatory lesions of the nasal mucous membranes, metaplasia of the olfactory epithelium and squamous epithelial hyperplasia of the forestomach was attributed to the irritant effect of the substance. There were no indications that monochloroacetic acid was carinogenic under these study conditions (NTP, 1990).

7.8 Reproductive toxicity

Pregnant Long-Evans rats were given monochloroacetic acid in water at daily oral doses of 0 (control), 17, 35, 70 or 140 mg/kg body weight from day 6 to day 15 of pregnancy. At the highest dose level, the body weight gain in the dams was significantly reduced compared with the control value, although organ weights were unaffected. The number of resorbed foetuses per litter and the weights of the live foetuses were comparable with the controls. The frequency of soft-tissue malformations ranged from 1.2% (controls) to 6.37% (140 mg/kg/day), but no dose-dependency was evident. Malformations of the cardiovascular system, particularly the left chambers of the heart, were significantly higher in the highest dose group than in the controls. No skeletal malformations were observed (Smith et al., 1990).

Monochloroacetic acid (analytical pure) dissolved in distilled water was intraperitoneally administered to 3 male mice (Swiss strain, 10–12 weeks old)/dose in doses of 12.5, 25 and 50 mg/kg body weight. 35 days after application sperms were examined. In the 25 and 50 mg/kg dose groups a significant increase in the number of abnormal sperms (giant head, dwarf type, almend shaped, amorphous type) was observed (Bhunya and Das, 1987).

7.9 Effects on the immune system

No information available.

7.10 Neurotoxicity

After acute oral administration of 320–380 mg monochloroacetic acid/kg to male Swiss-Webster mice, rigidity of the front paws was seen in 10% of the surviving animals. On histological examination of the brains of these mice 48 hours or more after treatment, erythrocytes were found outside the capillaries in various regions of the brain, particularly in the cerebellum. After 8 weeks, lysis of the extravascular erythrocytes occurred, with loss of Purkinje's cells in the cerebellum. 3 hours after acute oral administration of 380 mg/kg (LD_{80}), levels of intravenously administered ^{14}C-inulin and ^{3}H-dopamine (1.0 μCi) were significantly increased in all areas of the brain compared with the controls. This was interpreted as an impairment of the blood-brain barrier caused by monochloroacetic acid. The increase in ^{14}C-inulin in the brain after 2 hours was consistent with the toxic effects observed (Berardi et al., 1987).

Dichloroacetic acid and phenobarbital were tested in rodents as potential antidotes for monochloroacetic acid intoxication. No deaths occurred if rats were given dichloroacetic acid (110 mg/kg intraperitoneally) 15 minutes after the administration of monochloroacetic acid (80 mg/kg intravenously), while treatment with phenobarbital (no further details) reduced the mortality rate to less than 20%. A similar effect was also shown in mice. The author believed that the lethal effect of monochloroacetic acid in rats was not associated with altered permeability of the blood-brain barrier, because the administration of phenobarbital to rats that had been treated with a lethal dose of ^{14}C-monochloroacetic acid, did not change the concentration of monochloroacetic acid and its metabolites in the plasma or the cerebrospinal fluid. The lethal effect was more strongly associated with a dose-dependent accumulation of lactic acid in the cerebrospinal fluid, which occurred in parallel with the development of ataxia, and for which a threshold value for mortality (18 mmol/l) was observed. Dichloroacetic acid and phenobarbital were able to reduce the lactate level in the cerebrospinal fluid, and with this the mortality (Mitroka, 1990).

7.11 Other effects

Monochloroacetic acid inhibits acetate oxidation in vitro (0.5 ml [1-^{14}C]-acetate, 0.5 ml liver homogenate, 0.5 ml 0.5 M tris buffer (pH 7.4) and 1 ml 1 μM monochloroacetic acid; Hayes et al., 1973).

In vitro studies in rat liver slices have shown that monochloroacetic acid (0.1 mM) inhibits the incorporation of ^{14}C (no further details) into glucose. The production of ketone bodies and $^{14}CO_2$ was not affected. From data on the inhibitory effect of monochloroacetic acid (1 and 10 mM) on the fixation of CO_2 in mitochondria (measurement of carboxylase activity) it was concluded that gluconeogenesis is inhibited by monochloroacetic acid (Doedens and Ashmore, 1972).

In vitro, monochloroacetic acid (2.5–50 mM) reacts dose-dependently with glutathione transferase in rat liver by directly binding to it. A substrate of glutathione and 1-chloro-2,4-dinitrobenzene was used in this study. Binding of monochloroacetic acid to the various isoenzymes which were also studied was variable (Dierickx, 1984).

When male Sprague-Dawley rats (175 to 200 g) were given a single oral dose of 0.5 ml [1-^{14}C]-monochloroacetic acid (8.75 mg/kg body weight in 0.9% saline; 50 μCi, equivalent to 1.85×10^7 Bq), it was found to be conjugated to cholesterol in the liver after 24 hours (Bhat and Ansari, 1989).

Subcutaneous administration to rats of 162 mg/kg (LD_{90}) caused significant inhibition of the sulphydryl (SH) groups in the liver. The inhibition increased over a period of 120 minutes. Inhibition of SH-groups was also detected in the kidney cortex and medulla oblongata (Hayes et al., 1973).

Monochloroacetic acid has been detected as a metabolite after inhalation of trichloroethylene (Dancer et al., 1965), and possibly as the glutathione conjugate after oral administration of chloroethanol to rats (Grunow and Altmann, 1982). It was also found after administration of vinyl chloride (Plugge and Safe, 1977). In vitro, monochloroacetic acid was produced when 1,1,2-trichloroethane and 1,1,2,2-tetrachloroethane were incubated with liver microsomes. No covalent binding of monochloroacetic acid to components of the microsomes could be detected (Ivanetich and van Den Honert, 1981). Similarly, no alkylation of cysteine-SH took place in vitro (Hayes et al., 1973).

Sprague-Dawley rats and B6C3F$_1$ mice received monochloroacetic acid in their drinking water for 14 days at concentrations of 11 to 32 mM (1–3 g/l). In contrast to di- and trichloroacetic acid which were also studied, monochloroacetic acid did not induce liver enlargement or peroxisome proliferation in either species (DeAngelo et al., 1989).

In order to determine whether monochloroacetic acid increased the toxicity of vinylidene chloride, female Sprague-Dawley rats were given monochloroacetic acid (94 mg/kg body weight) by stomach tube one hour before an injection of vinylidene chloride (200 mg/kg body weight). Clearance experiments after 24 and 48 hours showed that pretreatment with monochloroacetic acid delayed recovery from adverse effects caused by vinylidene chloride, but it did not increase its toxicity, although monochloroacetic acid did reduce the glutathione level in the liver (Davis and Berndt, 1989.)

The protein content of HepG2 cells in culture was determined in a cytotoxicity test. The PJ_{50} value, which represents the concentration at which the protein content of the cells is reduced by 50% after 24 hours, was 10 mM for monochloroacetic acid (Dierickx, 1989).

8. Experience in humans

According to an old report, monochloroacetic acid vapours have produced corneal damage in humans (Knapp, 1923).

When monochloroacetic acid was absorbed through human skin (no details on quantities or the course of toxicity), there was a latent period of at least one hour before symptoms of reduced blood pressure, depressed breathing, spasms, vomiting and diarrhoea, bloody sputum and coma occurred, with deaths after 4 hours. Locally, first-degree burns developed. Post-mortem examination revealed bleeding of the internal organs, predominantly in the central nervous system, heart and kidneys (Zeldenrust, 1951).

A fatal poisoning was reported after accidental dermal exposure. Immediately after contamination with warm monochloroacetic acid solution (about 10% of the body surface), the skin was thoroughly rinsed with water for 1 hour, but death occurred 10 hours later. The symptoms progressed through first-degree burns, anxiety and considerable unrest to a deep state of shock (Christofano et al., 1970).

A 47-year-old worker was contaminated with pressurized molten monochloroacetic acid (ca. 90 °C) on both legs. Despite immediate rinsing with water, about 6% of the body surface suffered first degree burns. 4 hours after the accident, the worker was affected by nausea and vomiting, cardiovascular shock and neurological symptoms including lack of consciousness and finally coma. Heart arrhythmia, a drop in blood pressure and severe metabolic acidosis were observed. The patient was treated with ethanol, an effective antidote in the treatment of monofluoroacetic acid intoxication, via a gastric catheter. After 24 hours, the symptoms began to improve, and after 3 months the worker returned to employment. The authors suggested that ethanol could act as an acetate donor, and inhibit the introduction of chloroacetate into the citric acid cycle (Ruty et al., 1987).

A 38-year-old lorry driver was sprayed with an 80% solution of monochloroacetic acid while unloading. Despite rinsing with water for 20 minutes, 25 to 30% of the surface area of the body suffered first degree burns. After 1 hour, a rapid drop in blood pressure and excitation were observed, followed later by cardiovascular symptoms and unconsciousness as well as progressive kidney damage. Despite antidote therapy with ethanol and acetyl cysteine (as an SH-group donor) during the first 24 hours, the patient died on the 7th day after the accident (Kulling et al., 1986).

Other cases with fatal outcomes have been reported, in which death occurred within 6 to 18 hours of contact with either molten or concentrated solutions (80%) of monochloroacetic acid. In one case, the autopsy showed damage to the lungs and kidneys. Only one person, 5 to 10% of whose body surface was exposed to a concentrated solution of monochloroacetic acid, survived without antidote treatment (Millischer et al., 1987b).

In an accident in which a worker accidently spilt about 0.3 ml hot ^{14}C-labelled monochloroacetic acid (about 60–70 mc) over the fingers, a proportion of the monochloroacetic acid which landed on the skin was absorbed despite the hands being washed within a minute of the accident. Small blisters developed, but there were no signs of systemic toxicity. Details of the metabolic studies undertaken are given in the section on toxicokinetics and metabolism (7.1, p. 6; Dancer et al., 1965).

An English abstract of a Russian paper reports lung, heart and gastrointestinal diseases and disturbances in protein-lipid metabolism amongst workers engaged in the production of monochloroacetic acid and 2,4-dichlorophenoxyacetic acid (no exact details; Aliverdiyeva et al., 1983).

The threshold concentration for irritant effects in the eye (exposure for 1 minute) was given as an atmospheric concentration of 5.7 mg/m^3 (Maksimov and Dubinina, 1974; Rodionova and Ivanov, 1979).

The lethal oral dose in humans was estimated to be about 50 mg/kg (Gosselin et al., 1976).

Monochloroacetic acid has been described as a strongly irritant to corrosive material (Fassett, 1967; Toxic and Hazardous Industrial Chemicals Safety Manual, 1979; Watrous, 1974; Sax, 1981; Gosselin et al., 1976).

An irritant effect of monochloroacetic acid vapour on the eyes and respiratory tract has been mentioned (Watrous, 1974), but there are no descriptions in the literature of any corresponding symptoms.

9. Threshold limit values

Validated threshold values are not available (see Sect. 7.5).

References

Aliverdiyeva, S.S. et al.
Azerb. Med. ZH, 60 (2), 6 (1983)
Cited in: Toxicity Profile, Chloroacetic acid (CA)
BIBRA (The British Industrial Biological Research Association) (1988)

Amacher, D.E., Turner, G.N.
The detection of carcinogens from several chemical classes as mutagens in a modified L5178Y/TK mutation assay
Environ. Mutagen., 2, 254–255 (1980)

Amacher, D.E., Turner, G.N.
Mutagenic evaluation of carcinogens and non-carcinogens in the L5178Y/TK assay utilizing postmitochondrial fractions (S9) from normal rat liver
Mutat. Res., 97, 49–65 (1982)

Babanov, G.P., Verkhovskii, L.G., Shcherbakov, A.O.
Toxicological evaluation of sodium monochloroacetate
Deposited Doc., VINITI 4560-84, 20 pp. (1984)

Bartsch, H., Malavaille, C., Montesano, R.
Human, rat and mouse liver-mediated mutagenicity of vinyl chloride in *S. typhimurium* strains
Int. J. Cancer, 15, 429–437 (1975)

Bartsch, H., Malaveille, G., Camus, A.M., Martel-Planche, G., Brun, G.
Validation and comparative studies on 180 chemicals with *S. typhimurium* strains and V 79 Chinese hamster cells in the presence of various metabolizing systems
Mutat. Res., 76, 1–50 (1980)

Berardi, M., Snyder, R.
Toxicity and pharmacokinetics of monochloroacetic acid
Pharmacologist, 25, 228 (1983)

Berardi, M.R., Snyder, R., Waritz, R.S., Cooper, K.R.
Monochloroacetic acid toxicity in the mouse associated with blood-brain barrier damage
Fundam. Appl. Toxicol., 9, 469–479 (1987)

Bhat, H.K., Ahmed, A.E., Ansari, G.A.S.
Toxicokinetics of monochloroacetic acid: a whole-body autoradiography study
Toxicology, 63, 35–43 (1990)

Bhat, H.K., Ansari, G.A.S.
Covalent interaction of chloroacetic and acetic acids with cholesterol
J. Biochem. Toxicol., 4, 189–193 (1989)

Bhat, H.K., Kanz, M.F., Campbell, G.A., Ansari, G.A.S.
Ninety day toxicity study of chloroacetic acid in rats
Fundam. Appl. Toxicol., 17, 240–253 (1991)

Bhunya, S.P., Das, P.
Bone marrow chromosome aberration and sperm abnormality in mice in vivo induced by monochloroacetic acid (MCA)
Chromosome Inf. Serv., 42, 28–30 (1987)

Bionetics Research Labs., Inc.
Evaluation of carcinogenic, teratogenic and mutagenic activities of selected pesticides and industrial chemicals. Vol. I, Carcinogenic study prepared for National Cancer Institute
U.S. National Technical Information Services, report no. PB-223159 (1968)

Bryant, B.J., Jokinen, M.P., Eustis, S.L., Thompson, M.B., Abdo, K.M.
Toxicity of monochloracetic acid administered by gavage to F344 rats and B6C3F1 mice for up to 13 weeks
Toxicology, 72, 77–87 (1992)

Christiansen, M., Dalgaard-Mikkelsen, S.
Toxic effects of monochloroacetate on geese
Acta Pharmacol. Toxicol., 18, 179–182 (1961)

Christofano, E.E., Frawley, J.P., Reed, H.L., Keplinger, M.L.
Skin exposure to monochloroacetic acid
Am. Ind. Hyg. Assoc. J., 19, 35 (1970)

Dalgaard-Mikkelsen, S., Kvorning, S.A., Møller, K.O.
Toxic effects of monobromoacetic acid on pigs
Acta Pharmacol. Toxicol., 11, 13–32 (1955)

Dalgaard-Mikkelsen, S., Rasmussen, F.
Toksiciteten af monokloracetat for kvaeg
Nord. Vet. Med., 13, 271–279 (1961)

Dancer, G.H., Morgan, A., Hutchinson, W.P.
A case of skin contamination with carbon-14 labelled chloroacetic acid
Health Phys., 11, 1055–1058 (1965)

Daniel, F.B., Robinson, M., Stober, J.A., Page, N.P., Olson, G.R.
Ninety-day toxicity study of sodium monochloroacetate in Sprague-Dawley rats
Toxicology, 67, 171–185 (1991)

Davis, M.E., Berndt, W.D.
Acute toxicity of monochloroacetate in male and female rats
Fed. Proc., 46 (3), 957 (1987)
71st Annual Meeting of the Federation of American Societies for Experimental Biology, Washington, 29. 03. 1987

Davis, M.E., Berndt, W.D.
Monochloroacetate (MCA) does not enhance vinylidene chloride (VDC) toxicity in female rats
Fed. Am. Soc. Exp. Biol. J., 3, A1038 (1989)

DeAngelo, A.B., Daniel, F.B., McMillan, L., Wernsing, P., Savage, R.E.
Species and strain sensitivity to the induction of peroxisome proliferation by chloro acetic acids
Toxicol. Appl. Pharmacol., 101, 285–298 (1989)

Dierickx, P.J.
In vitro binding of acetic acid and its chlorinated derivatives by the soluble glutathione S-transferases from the rat liver
Res. Commun. Chem. Pathol. Pharmacol., 44, 327–330 (1984)

Dierickx, P.J.
Cytotoxicity testing of 114 compounds by the determination of the protein content in HepG2 cell cultures
Toxic. in Vitro, 3, 189–193 (1989)

Doedens, D., Ashmore, J.
Inhibition of pyruvate carboxylase by chloropyruvic acid and related compounds
Biochem. Pharmacol., 21, 1745–1751 (1972)

Duuren, van, B.L., Goldschmidt, B.M., Katz, C., Seidman, I., Paul, J.S.
Carcinogenic activity of alkylating agents
J. Natl. Cancer Inst., 53, 695–700 (1974)

Emerson, G., Morrison, I.
Fed. Proc., 2, 78 (1943)
Cited in: Dalgaard-Mikkelson et al. (1955)

Fassett, D.W.
Organic acids, anhydrides, lactones, acid halides and amides, thioacids
In: Industrial hygiene and toxicology
Patty, F.A. (ed.), vol. 2, p. 1795–1796
Interscience Publishers, New York (1967)

Fuhrmann, F.A., Field, J., Wilson, R.H., Deeds, F.
Monochloroacetate: effects of chronic administration to rats on growth, activity and tissue metabolism and inhibitory effects in vitro compared with monoiodoacetate and monobromoacetate
Arch. Int. Pharmacodyn., 102, 113–125 (1955)

Galloway, S.M., Armstrong, M.J., Reuben, C., Colman, S., Brown, B., Cannon, C., Bloom, A.D., Nakamura, F., Ahmed, M., Duk, S., Rimpo, J., Margolin, B.H., Resnick, M., Anderson, B., Zeiger, E.
Chromosome abberations and sister chromatid exchanges in Chinese hamster ovary cells. Evaluation of 108 chemicals
Environ. Mol. Mutagen., 10 (10), 1–175 (1987)

Gosselin, R.E., Hodge, H.C., Smith, R.P., Gleason, M.N.
Clinical toxicology of commercial products, acute poisoning
4th ed., p. 133
The Williams and Wilkins Co., Baltimore (1976)

Gross, B.J., Branchflower, R.V., Burke, T.R., Lees, D.E., Pohl, L.R.
Bone marrow toxicity in vitro of chloroamphenicol and its metabolites
Toxicol. Appl. Pharmacol., 64, 557–565 (1982)

Grunow, W., Altmann, H.J.
Toxicokinetics of chloroethanol in the rat after single oral administration
Arch. Toxicol., 49, 275–284 (1982)

Hayes, F.D., Gehring, P.J., Gibson, J.E.
Studies on the acute toxicity of monochloroacetic acid in rats
Toxicol. Appl. Pharmacol., 22, 303, Abstr. 76 (1972)

Hayes, F.D., Short, R.D., Gibson, J.E.
Differential toxicity of monochloroacetate, monofluoroacetate and monoiodoacetate in rats
Toxicol. Appl. Pharmacol., 26, 93–102 (1973)

Hoechst AG, Pharma Forschung Toxikologie
Akute orale Toxizität von Monochloressigsäure VA 2308 an weiblichen Ratten
Unpublished report no. 232/79 (1979 a)

Hoechst AG, Pharma Forschung Toxikologie
Akute dermale Toxizität von Monochloressigsäure VA 2308 an weiblichen Ratten
Unpublished report no. 234/79 (1979 b)

Hoechst AG, Pharma Forschung Toxikologie
Akute dermale Toxizität von Monochloressigsäure VA 2308 an Kaninchen
Unpublished report no. 236/79 (1979 c)

Hoechst AG, Pharma Forschung Toxikologie
Akute subcutane Toxizität von Monochloressigsäure VA 2308 an weiblichen Ratten
Unpublished report no. 233/79 (1979 d)

Hoechst AG, Pharma Forschung Toxikologie
Haut- und Schleimhautverträglichkeit von Monochloressigsäure VA 2308 an Kaninchen
Unpublished report no. 235/79 (1979 e)

Hoechst AG, Pharma Forschung Toxikologie
Ames-Test, Substanz: 055/79 Monochloressigsäure
Unpublished report no. 474/79 A (1979 f)

Hoechst AG
Grunddatensatz Monochloressigsäure (1991)

Huberman, E., Bartsch, H., Sachs, L.
Mutation induction in Chinese hamster V 79 cells by two vinyl chloride metabolites, chloroethylene oxide and 2-chloro-acetaldehyde
Int. J. Cancer, 16, 639–644 (1975)

Innes, J.R.M., Ulland, B.M., Valerio, M.G., Petrucelli, L., Fishbein, L., Hart, E.R., Pallotta, A.J., Bates, R.R., Falk, H.L., Gart, J.J., Klein, M., Mitchell, I., Peters, J.
Bioassay of pesticides and industrial chemicals for tumorigenicity in mice: a preliminary note
J. Natl. Cancer Inst., 42, 1101–1114 (1969)

Ivanetich, K.M., van den Honert, L.H.
Chloroethanes: their metabolism by hepatic cytochrome P-450 in vitro
Carcinogenesis, 2, 697–702 (1981)

Knapp, P.
Schweiz. Wochenschr., 4, 702 (1923)
Cited in: Toxicity Profile, Chloroacetic acid (CA)
BIBRA (The British Industrial Biological Research Association) (1988)

Koenig, G.
Chloressigsäuren
In: Ullmanns Encyklopädie der technischen Chemie
4th ed., vol. 9, p. 393–397
Verlag Chemie, Weinheim (1975)

Kulling, P., Andersson, H., Böström, K., Johansson, L.A., Lindström, B., Nyström, B.
Lethal systemic poisoning after skin exposure to monochloroacetic acid
III World Congress of the World Federation of Associations of Clinical Toxicology and Poison Control Centres and XII International Congress of the European Association of Poison Control Centres (EAPCC), Brüssel, 27.–30. August 1986

Kurĉatov, G.V., Vasileva, Z.A.
Study of thiol compounds as possible antidotes to poisoning with ethylene chlorohydrin and monochloroacetic acid
Fiziologiĉeski Aktivnye Veŝĉestva, 8, 55–58 (1976)

Maksimov, G.G., Dubinina, O.N.
Empirical determination of the MAC value for chloroacetic acid in production plant atmospheres (English translation)
Gig. Tr. Prof. Zabol., 9, 32–35 (1974)

Malaveille, C., Bartsch, H., Barbin, A., Camus, A.M., Montesano, R., Croisy, A., Jacquignon, P.
Mutagenicity of vinyl chloride, chloroethylene oxide, chloroacetaldehyde and chloroethanol
Biochem. Biophys. Res. Commun., 63, 363–370 (1975)

Mamber, S.W., Bryson, V., Katz, S.E.
The Escherichia coli WP2/WP100 assay for detection of potential chemical carcinogens
Mutat. Res., 119, 135–144 (1983)

McCann, J., Choi, E., Yamasaki, E., Ames, B.N.
Detection of carcinogens as mutagens in the Salmonella/microsome test: assay of 300 chemicals
Proc. Natl. Acad. Sci. USA, 72, 5135–5139 (1975)

McGregor, D.B., Martin, R., Cattanach, P., Edwards, I., McBride, D., Caspary, W.J.
Responses of the L5178Ytk^{+}/tk^{-} mouse lymphoma cell forward mutation assay to coded chemicals. I. Results for nine compounds
Environ. Mutagen., 9, 143–160 (1987)

Millischer, R.J., Gonnet, J.F., Ruty, J., Vincenti, M., Jouglard, J., Contassot, J.C.
Monochloroacetic acid (MCA): evidence of systemic toxicity from percutaneous contact and study of ethanol as an antidote in animals
Occup. Health in the Chem. Ind., 22th ICOH-Congress, Sydney (1987a)

Millischer, R.J., Jouglard, J., Vincenti, M., Ruty, J., Contassot, J.C.
Monochloroacetic acid: seven worldwide cases of systemic poisoning resulting from accidental skin contact
Occup. Health in the Chem. Ind., 22th ICOH-Congress, Sydney (1987b)

Mitroka, J.G.
Monochloroacetic acid lethality in the rat in relation to lactic acid accumulation in the cerebrospinal fluid
Diss. Abstr. Int., 50, 5000-B (1990)

Morrison, J.L.
Toxicity of certain halogen substituted aliphatic acids for white mice
J. Pharmacol. Exp. Ther., 86, 336–338 (1946)

Mortelmans, K., Haworth, S., Lawlor, T., Speck, W., Tainer, B., Zeiger, E.
Salmonella mutagenicity tests. II. Results from the testing of 270 chemicals
Environ. Mutagen., 8 (Suppl. 7), 1–119 (1986)

Nakamura, S., Oda, Y., Shimada, T., Oki, I., Sugimoto, K.
SOS-inducing activity of chemical carcinogens and mutagens in Salmonella typhimurium TA1535/pSK 1002: examination with 151 chemicals
Mutat. Res., 192, 239–246 (1987)

NTP (National Toxicology Program)
Draft of the NTP Technical Report on the toxicology and cancerogenesis studies of monochloroacetic acid (CAS-No. 79-11-8) in F344/N rats and B6C3F1 mice (gavage studies)
NTP TR 396, NIH Publication No. 90-2851 (1990)

Ono, Y., Somiya, I., Kawamura, M.
The evaluation of genotoxicity using DNA repairing test for chemicals produced in chlorination of ozonation processes
Wat. Sci. Tech., 23, 329–338 (1991)

Plugge, H., Safe, S.
Vinylchloride metabolism – a review
Chemosphere, 6, 309–325 (1977)

Quick, M.P., Manser, P.A., Stevens, H., Bolton, J.F.
Sodium monochloroacetate poisoning of cattle and sheep
Vet. Rec., 155–156 (1983)

Rannug, U., Göthe, R., Wachtmeister, C.A.
The mutagenicity of chloroethylene oxide, chloroacetaldehyde, 2-chloroethanol and chloroacetic acid, conceivable metabolites of vinylchloride
Chem. Biol. Interact., 12, 251–263 (1976)

Rodionova, R.P., Ivanov, N.G.
Comparison of the level of irritant properties of industrial toxins
Toksikologiya Novykh Promyshlennykh Khimichesrikh Veshchestv., 15, 58–63, 145–150 (1979)

Ruty, J., Millischer, R.J., Contassot, J.C., Vincenti, M., Jouglard, J.
Monochloroacetic acid: a report of systemic poisoning from percutaneous absorption
Occup. Health in the Chem. Ind., 22th ICOH-Congress, Sydney (1987)

Sawada, M., Sofuni, T., Ishidate, M.
Cytogenetic studies on 1,1-dichloroethylene and its two isomers in mammalian cells in vitro and in vivo
Mutat. Res., 187, 157–163 (1987)

Sax, N.I. (ed.)
Dangerous properties of industrial materials
5th ed., p. 838
van Nostrand Reinhold Company, New York (1981)

Smith, M.K., Randall, J.L., Read, E.J., Stober, J.A.
Developmental effects of chloroacetic acid in the Long-Evans rat
Teratology, 41, 593 (1990)

Sonnenfeld, G., Barnes, M.C., Schooler, J., Streips, U.N.
Inhibition of interferon induction as a screen for the carcinogenic potential of chemicals
Interferon: properties and clinical uses
Proc. Int. Symp., 589–598 (1980)

Szegedi, M.
Genotoxic activities of 3-chloropropionic acid and related compounds
Environ. Mol. Mutagen., 14, Suppl. 15, 196 (1989)

Toxic and Hazardous Industrial Chemicals Safety Manual
The International Technical Information Institute, Tokyo, p. 116–117 (1979)

University of Wisconsin-Madison
Cited in: NTP (1990)

Watrous, R.M.
Health hazards of the pharmaceutical industry
Br. J. Ind. Med., 4, 111–125 (1947)

Woodard, G., Lange, S.W., Nelson, K.W., Calvery, H.O.
The acute oral toxicity of acetic, chloroacetic, dichloroacetic and trichloroacetic acids
J. Ind. Hyg. Toxicol., 23, 78–82 (1941)

Yllner, S.
Metabolism of chloroacetate-1-^{14}C in the mouse
Acta Pharmacol. Toxicol., 30, 69–80 (1971)

Zeldenrust, J.
Over een geval van peracute vergiftiging door monochloorazijnzuur
Arch. Belg. Med. Soc., 9, 19–20 (1951)

Benzene sulphonic acid hydrazide

1. Summary and assessment

Benzene sulphonic acid hydrazide is acutely toxic (oral LD_{50} rat: 160 mg/kg body weight).

Mice, rats and rabbits tolerate single 1-hour und 4-hour periods inhaling the decomposition products of benzene sulphonic acid hydrazide at 110 °C without symptoms occurring.

Benzene sulphonic acid hydrazide is non-irritant to the skin and eyes of the rabbit.

Benzene sulphonic acid hydrazide is point mutagenic in the Salmonella/microsome test in the strains TA 1535 and TA 100, both with and without metabolic activation, but is not point mutagenic in the strains TA 1537 and TA 98. However, the effective doses are relatively high.

According to one relatively old, inadequately conducted and reported study, a marginal tumorigenic effect of benzene sulphonic acid hydrazide cannot be ruled out.

Studies on the reproductive toxicity of benzene sulphonic acid hydrazide (75% in paraffin oil) are available only for chicken embryos. Doses lethal to the embryos lead to defects in the eyes and the wall of the coelom, as well as to oedema formation. However, it is almost impossible to apply these findings to mammals since the absorption barriers and detoxification mechanisms in the latter are not taken into account.

Experience in humans is limited to a short report of an uncertain case of skin-sensitization in a 63-year old employee in the rubber industry.

2. Name of substance

2.1	Usual name	Benzene sulphonic acid hydrazide
2.2	IUPAC-name	Benzene sulphonic acid hydrazide
2.3	CAS-No.	80-17-1

3. Synonyms, common and trade names

Benzene sulfonic acid hydrazide
Benzenesulphonyl hydrazide
Benzenesulphohydrazide
BSH
Benzenesulphonylhydrazine
Porofor BSH
Benzolsulfonsäurehydrazid

4. Structural and molecular formulae

4.1 Structural formula

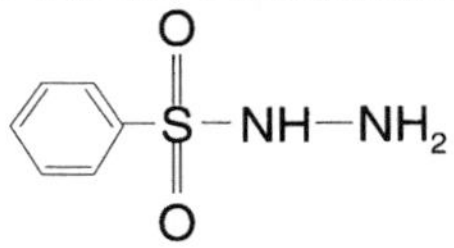

4.2 Molecular formula $C_6H_8O_2N_2S$

5. Physical and chemical properties

5.1	Molecular mass, g/mol	172.21
5.2	Melting point, °C	102–103 starts decomposing at 70 °C (Eichenhofer and Roos, 1977; Bayer, 1988)
5.3	Boiling point, °C	no information available
5.4	Vapour pressure, hPa	no information available
5.5	Density, g/cm³	1.48 (Eichenhofer and Roos, 1977)
5.6	Solubility in water	15 g/l (at 20 °C) (Bayer, 1988)
5.7	Solubility in organic solvents	soluble in dimethyl sulphoxide, ethanol (Bayer, 1988)

5.8 Solubility in fat	no information available
5.9 pH-value	neutral (Bayer, 1988)
5.10 Conversion factor	1 ppm ≙ 7.03 mg/m^3 1 mg/m^3 ≙ 0.14 ppm (at 25 °C and 1013 hPa)

6. Uses

Blowing agent in the manufacture of cellular rubber and sponge rubber, form soles, hard-cell soles and phenolic resin foams (Eichendorfer and Roos, 1977).

7. Experimental results

7.1 Toxicokinetics and metabolism

No information available.

7.2 Acute and subacute toxicity

Following a 14-day post-treatment observation period, oral administration by gavage of a single dose of benzene sulphonic acid hydrazide in polyethylene glycol 400 yielded an LD_{50} of 160 (140 to 190) mg/kg body weight for male Wistar rats, 160–180 g in weight (10 animals/dose group). While a dose of 50 mg/kg body weight did not produce any symptoms of intoxication, all animals treated with ≥100 mg/kg body weight exhibited tremors, clonical spasms, jumping and aggressiveness, as well as poor general condition (Bayer, 1976).

In order to determine the acute inhalation toxicity, 5 male Wistar rats (180 to 210 g), 10 male NMRI mice (25 to 35 g) and 2 to 3 Chinchilla rabbits (3050 to 3400 g) of both sexes were used for each concentration and exposure time, and were exposed for 1 and 4 hours to the products resulting from decomposition of benzene sulphonic acid hydrazide at 110 °C. The post-treatment observation period was 7 days. Single 1-hour inhalation of the vapour at a concentration of 5055 mg/m^3 was tolerated by rats and mice without symptoms; rabbits tolerated single 1-hour exposure to 2900 mg/m^3 without symptoms. Even after 4 hours exposure, none of the rats and mice that had inhaled 5475 mg/m^3 or the rabbits that had inhaled 4332 mg/m^3 died during the post-treatment observation period. Body weight gain in the animals was normal. The LC_{50} values for 4-hour exposure to the decomposition products of benzene sulphonic acid hydrazide were thus greater than 5475 mg/m^3 for rats and mice, and greater than 4332 mg/m^3 for rabbits (Bayer, 1978).

7.3 Skin and mucous membrane effects

500 mg/animal was applied to the inner surface of the ear of white male and female New Zealand rabbits (3–4 kg), the test article being attached for 24 hours using an adhesive dressing. The post-treatment observation period was 7 days. The treatment produced no skin changes of any kind. Benzene sulphonic acid hydrazide was assessed as non-irritant (irritant value 0; Bayer, 1977).

Nor was there any evidence of an irritant effect on the eye during the 7-day post-treatment observation period following administration of 50 mg into the conjunctival sac of 2 rabbits of each sex (irritant value 0; Bayer, 1977).

7.4 Sensitization

No information available.

7.5 Subchronic and chronic toxicity

No information available.

7.6 Genotoxicity

7.6.1 In vitro

Benzene sulphonic acid hydrazide was tested for mutagenic effect in the Salmonella/microsome test in strains TA 1535, TA 100, TA 1537 and TA 98 with and without the addition of the S9-fraction from the liver of male Sprague-Dawley rats pretreated with Aroclor 1254. Concentrations up to 12 500 µg/plate were tested. The substance was positive in strains TA 1535 and TA 100, both with and without activation. 625 µg/plate was given as the lowest effective dose for TA 1535 and 5000 µg/plate for TA 100 (Bayer, 1983).

7.6.2 In vivo

No information available.

7.7 Carcinogenicity

30 female Swiss mice (8 weeks old at the start of the study) received 2 mg benzene sulphonic acid hydrazide intragastrally in 0.2 ml 3% aqueous gelatine solution 5 times/week for 4 weeks, then 3 times/week up to a total number of 158 doses. An equally sized control group received 0.2 ml of the 3% gelatine solution according to the same scheme. After 60 weeks, the surviving animals were examined for tumours. Among 24 mice there were 4 lung tumours (controls 1/29) and 4 liver tumours (controls 1/29). No further details of the distribution of the tumours and the histopathological findings were available. The authors concluded from this result that benzene sulphonic acid hydrazide may have marginal tumorigenic activity in mice, but is not a strong carcinogen (Cremlyn and Roe, 1971).

7.8 Reproductive toxicity

In order to study the embryotoxicity, benzene sulphonic acid hydrazide (75% Porofor BSH in paraffin oil) dissolved in acetone was administered at concentrations of 1.0, 2.0, 3.0 and 4.0 µmol/egg into the air chamber of 3 day old chicken embryos (white Leghorn). The injection volume was 1 µl. 30 eggs were used for each concentration. A control consisting of 10 eggs received 5 µl acetone/egg. The eggs were examined against light at 2–3 day intervals, those containing dead embryos were sorted out and opened in order to check the developmental stage (according to the evaluation scale by Hamburger and Hamilton, 1951) and any external malformations. At the end of the 14-day incubation period (11 days after administration), the embryotoxic potency was evaluated according to the total number of embryos affected 11 days after administration, using overall mortality and the number of malformed embryos (total effect) as a basis. At 1.0 µmol, 13% of the treated eggs were affected, at 2.0 and 3.0 µmol 47% and at 4.0 µmol this figure was 70%. From these values, an ED_{50} (total effect) of 2.6 µmol/egg (confidence range 2.1 to 3.3 µmol) was determined over the 11 days following application. The malformations induced in all the dose groups were described as 2 different types of eye defect (reduced eye cup and partial or complete absence of lids and cornea), defects of the coelom wall and generalized oedema. The highest mortality rate (46%) and malformation rate (43%) determined occurred in the (dead and surviving) embryos treated with 4.0 µmol/egg (Korhonen et al., 1982, 1983). It is, however, almost impossible to apply these findings to mammalian systems since the absorption barriers and detoxification mechanisms in the latter are not taken into account (Skofitsch, 1988).

7.9 Effects on the immune system

No information available.

7.10 Neurotoxicity

No information available.

7.11 Other effects

No information available.

8. Experience in humans

Over a period of 6 months recurrent and occupationally-related vesicular eczema with localization on the surfaces of the hand, including the fingers, were observed in a 63-year old employee who had worked for 35 years in the rubber industry. After testing rubber specimens from the standard European series, as well as 8 different non-vulcanized rubber specimens using the patch test, a positive response was established on the 2nd and 4th day with several of the rubber specimens, although these did not belong to the standard range. Subsequent testing of individual components of the corresponding rubber specimens produced a marked vesicular response in the patient to 1% benzene sulphonic acid hydrazide in Vaseline. 25 additional control substances were negative (English and White, 1986).

References

Bayer AG, Institut für Toxikologie
Akute orale Toxizität, Porofor BSH-Pulver
Unpublished studies (1976)

Bayer AG, Institut für Toxikologie
Untersuchung zur Haut- und Schleimhautverträglichkeit, Porofor BSH
Unpublished studies (1977)

Bayer AG, Institut für Toxikologie
Porofor BSH, Bestimmung der Inhalationstoxizität (Zersetzungsdämpfe)
Unpublished studies (1978)

Bayer AG, Institut für Toxikologie
Porofor BSH, Salmonella/Microsomen-Test zur Untersuchung auf punktmutagene Wirkung
Unpublished report No. 11929 (1983)

Bayer AG, Geschäftsbereich Kautschuk
DIN Safety Sheet Porofor BSH Paste (1988)

Cremlyn, R.J.W., Roe, F.J.C.
A study of certain substituted sulfonohydrazides for carcinogenicity in mice
Food Cosmet. Toxicol., 9, 319–321 (1971)

Eichenhofer, K.W., Roos, E.
Hydrazin
In: Ullmanns Encyklopädie der technischen Chemie
4th ed., vol. 13, p. 95–107
Verlag Chemie GmbH, Weinheim (1977)

English, J.S.C., White, I.R.
Contact dermatitis from benzene sulphonyl hydrazide
Contact Dermatitis, 14, 183 (1986)

Hamburger, V., Hamilton, H.L.
A series of normal stages in the development of the chick embryo
J. Morphol., 88, 49–92 (1951)
Cited in: Korhonen et al. (1982)

Korhonen, A., Hemminki, K., Vainio, H.
Embryotoxicity of benzothiazoles, benzenesulfohydrazide and dithiodimorpholine to the chicken embryo
Arch. Environ. Contam. Toxicol., 11, 753–759 (1982)

Korhonen, A., Hemminki, K., Vainio, H.
Toxicity of rubber chemicals towards three-day chicken embryos
Scand. J. Work Environ. Health, 9, 115–119 (1983)

Skofitsch, G.
Vom Hund zum Ei? Große und kleine Laboratoriumstiere
In: Lembeck, F. (ed.), Alternativen zum Tierversuch
Georg Thieme Verlag Stuttgart, New York, 71–79 (1988)

1,3-Di-o-tolylguanidine

1. Summary and assessment

1,3-Di-o-tolylguanidine is acutely toxic (oral LD_{50} rat: 134 mg/kg). In guinea-pigs and rabbits, oral administration of sub-lethal doses (no further details) repeated up to 12 times yield evidence of liver and kidney lesions.

1,3-Di-o-tolylguanidine is non-irritant to the skin of the rabbit but has an irritant effect on rabbits eyes.

1,3-Di-o-tolylguanidine is non-point-mutagenic in the Salmonella/microsome test.

Reproductive toxicity studies of 1,3-di-o-tolylguanidine only exist for chicken embryos. In the dose range lethal to chicken embryos, defects in the eyes and wall of the coelom as well as wing abnormalities occur. It is, however, almost impossible to apply these findings to mammals since their characteristic absorption barriers and detoxication mechanisms are not taken into account in these studies.

1,3-Di-o-tolylguanidine ist stored preferentially in the sigma receptors in the central nervous system, and produces posture and movement stereotypy in rats when injected intracerebrally.

2. Name of substance

2.1	Usual name	1,3-Di-o-tolylguanidine
2.2	IUPAC-name	N,N'-Bis(2-methylphenyl)-guanidine
2.3	CAS-No.	97-39-2

3. Synonyms, common and trade names

1,3-Bis(o-tolyl)guanidine
Di-o-tolylguanidin
Di-o-tolylguanidine
1,3-Di-o-tolylguanidine

N,N'-Di-o-tolylguanidine
1,3-Di(2-tolyl)guanidine
DOTG
Soxinol DT
Vulcafor DOTG
Vulkacit DOTG

4. Structural and molecular formulae

4.1 Structural formula

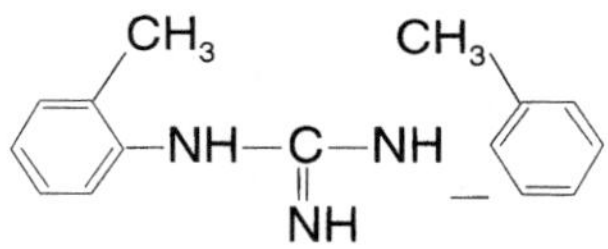

4.2 Molecular formula $C_{15}H_{17}N_3$

5. Physical and chemical properties

5.1	Molecular mass, g/mol	239.35
5.2	Melting point, °C	179 (Doehlemann, 1976)
5.3	Boiling point, °C	no information available
5.4	Vapour pressure, hPa	no information available
5.5	Density, g/cm³	1.18 (Doehlemann, 1976)
5.6	Solubility in water	insoluble (Doehlemann, 1976)
5.7	Solubility in organic solvents	soluble in acetone, ethanol, ethyl acetate, dichloromethane, slightly soluble in benzene, insoluble in benzine and carbon tetrachloride (Doehlemann, 1976)
5.8	Solubility in fat	no information available
5.9	pH-value	no information available

5.10 Conversion factor 1 ppm ≙ 9.77 mg/m³
1 mg/m³ ≙ 0.10 ppm
(at 25 °C and 1013 hPa)

6. Uses

Vulcanisation accelerator (Doehlemann, 1976).

7. Experimental results

7.1 Toxicokinetics and metabolism

No information available.

7.2 Acute und subacute toxicity

In order to test the acute toxicity of 1,3-di-o-tolylguanidine 103 to 216 mg/kg body weight was given as a 2% (w/v) suspension in propylene glycol once by gavage to groups of 5 male and 5 female Wistar rats (90 to 145 g and 89 to 116 g, respectively). The post-treatment observation period was 14 days. The LD_{50} was calculated at 134 (110 to 164) mg/kg body weight. Lethargy, tremors, and spasms occurred within a few hours of administration. Rats given the high doses died within 21 hours, the ones that survived recovered and were in good general condition at the end of the post-treatment observation period. Necropsy revealed no substance-related pathological changes (CIVO/TNO, 1975).

An oral LD_{50} of 500 mg/kg was obtained for 1,3-di-o-tolylguanidine in 6 rats (Dieke et al., 1947).

A value of 120 mg/kg was obtained as the minimum lethal oral dose for 1,3-di-o-tolylguanidine (technical grade, as hydrochloride) in guinea-pigs and rabbits. The animals died after 45 minutes. Higher doses were lethal within 12 minutes. The symptoms observed were signs of exhaustion, muscular twitching, dyspnoea and cyanosis (no further details; Smyth, 1931).

In a further study, the minimum lethal dose for rabbits was given as 80 mg/kg body weight (no further details; NIOSH, 1991).

In order to determine the intraperitoneal LD_{50}, small groups of CF1-mice (20 to 25 g; no details of numbers) were given increasing doses of 1,3-di-o-tolylguanidine intraperitoneally. The LD_{50} determined after 7 days was 25 to 50 mg/kg. Symptoms of intoxication were not reported (Doull et al., 1962).

With guinea-pigs and rabbits, repeated daily oral administration of sublethal doses (no further details) of 1,3-di-o-tolylguanidine for up to 12 days led to slight albuminuria, but not to urinary casts, and to the temporary appearance of reducing substances in the urine. Necropsy revealed evidence of liver and kidney lesions. There was apparently no marked cumulative effect (no further details; Smyth, 1931).

7.3 Skin and mucous membrane effects

In order to test the primary skin irritant effect, 500 mg of 1,3-di-o-tolylguanidine was applied beneath a patch to the intact and scarified skin of each of 6 adult New Zealand rabbits and fixed with an adhesive dressing for 24 hours. Observations and assessments were made 24 and 72 hours after application using the Draize method (1944). There was no evidence of a skin irritant effect due to 1,3-di-o-tolylguanidine either 24 hours or 72 hours after application (CIVO/TNO, 1975).

In order to determine the irritant effect on the eyes, 100 mg of the substance was applied to the conjunctival sac of one eye of each of 6 New Zealand rabbits. The eyes were examined 24, 48, 72 hours and 7 days after application and assessed using the methods prescribed by the FDA (1963) and Draize and Kelley (1952). 24 hours after application 3 of the 6 rabbits exhibited slight lesions to the cornea, and 4 out of 6 slight iritis; there was also slight to moderate conjunctivitis in all the rabbits. The changes in the eye receded gradually, but not completely, so that after 7 days it was still possible to observe slight conjunctivitis in half of the animals. On the basis of these results 1,3-di-o-tolylguanidine was assessed as irritant (CIVO/TNO, 1975).

7.4 Sensitization

No information available.

7.5 Subchronic and chronic toxicity

No information available.

7.6 Genotoxicity

7.6.1 In vitro

1,3-Di-o-tolylguanidine was tested for genotoxic effects in the Salmonella/microsome test in strains TA 1535, TA 1537, TA 1538, TA 98 and TA 100 with and without the addition of S9-fraction from the livers of male Sprague-Dawley rats pretreated with Aroclor 1254. The results were negative (no details of concentrations). Even at the higher concentrations (no details) no toxic effect was observed, and the result was thus regarded by the authors as inconclusive (Rannug et al., 1984).

No point mutagenic effect was observed in a further test on strains TA 98, TA 100, TA 1535, TA 1537, TA 1538 with and without S9-mix and using doses of 10, 30, 100, 300, 1000 and 3000 µg/plate (Sumitomo, 1981).

7.6.2 In vivo

No information available.

7.7 Carcinogenicity

No information available.

7.8 Reproductive toxicity

The effect of 1,3-di-o-tolylguanidine (technical grade) on the development of 3-day old chicken embryos (white Leghorn) was investigated. Concentrations of 0.05, 0.10, 0.20 and 0.40 µmol/egg dissolved in acetone were administered at a volume of 5 µl into the air chamber via a window which had been cut into the eggs after 3 days incubation (72–76 hours). The window was sealed with a transparent adhesive strip. 30 eggs were used for each concentration, 20 for the highest dose. 10 eggs treated with 5 µl acetone only served as controls. The eggs were examined against light at intervals of 2 to 3 days, and those containing dead embryos opened in order to check the developmental stage (according to the evaluation scale of Hamburger and Hamilton, 1951) and any external malformations. At the end of the 14-day incubation period (11 days after administration), the numbers of surviving embryos and external malformations were determined. The embryotoxic potency was evaluated using the total number of embryos affected 11 days after administration, using overall mortality and number of malformed embryos as a basis. At 0.05 and 0.10 µmol, 10% of the treated embryos were affected, at 0.20 µmol 67% and at 0.40 µmol the figure was 100%. An ED_{50} of 0.17 (0.12 to 0.22) µmol/egg was therefore obtained over a period of 11 days following administration and an LD_{50} estimated at 0.2 µmol/egg. Since even in the early stage of development (within 2 days of administration) the highest dose proved lethal to all embryos, it was not possible to observe any malformations in this group. The highest malformation rate of 13% was obtained with 0.2 µmol/egg. With regard to malformations, 2 different types of eye defect were observed (eye cup reduced in size and partial or complete absence of lid and cornea), as well as wing anomalies and defects in the coelom wall. On the basis of these findings, of 8 compounds, 1,3-di-o-tolylguanidine was assessed by the authors as the strongest acting after acrolein (Korhonen et al., 1983 a, 1983 b). It is, however, almost impossible to apply these results to mammals since the absorption barriers and detoxification mechanisms characteristic of mammals are not taken into account (Skofitsch, 1988).

7.9 Effects on the immune system

No information available.

7.10 Neurotoxicity

1,3-Di-o-tolylguanidine accumulates preferentially in the sigma-receptors in the brain. Its effect on the behaviour and coordinated movements of 21 male Sprague-Dawley rats was thus investigated. The substance (dissolved in acetic acid and adjusted to pH 7 with NaOH and distilled water) was administered intracerebroventricularly at a volume of 5 µl/rat (no details of the concentration injected). 5 minutes after treatment, peculiarities in behaviour, such as sniffing, standing erect, circus and backward movements as well as to and fro movements of the head could be observed. The dose which caused this stereotyped behaviour in 50% of the animals (ED_{50}) was given as 55 (45 to 67) nmol/rat. 57 (47 to 70) nmol/rat was obtained as the ED_{50} for induction of ataxia (Contreras et al., 1988).

The effect which substances with a high affinity for sigma-receptors have on the motor centres of the brain in which sigma receptors are present was also investigated. 17 rats were given 1,3-di-o-tolylguanidine (in the form of its acetate in aqueous solution) at a volume of 0.5 µl and doses of 1.5, 3.7 and 9.3 nmol by means of a canula implanted in the red nucleus. The dose was administered within 72 seconds. Within 30 minutes of the injection the authors observed increasing dystonia and torticollis. Microinjections of 0.07, 0.29, 1.15, 4.65 and 18.6 nmol into the substantia nigra caused dose-dependent increasing circus movements in 33 rats (even at the extremly low dose of 0.29 nmol). Posture and motor response were generally normal again within 90 minutes. Microscopic evaluation of the brain in the animals affected by torticollis revealed no pathological changes. The authors concluded from these results that the effects were attributable to pharmacological effects and not to structural changes in tissue, and that sigma receptors play a role in the pathogenesis of dystonia (Walker et al., 1987, 1988).

1,3-Di-o-tolylguanidine had an inhibiting effect on the carbachol-stimulated phosphoinositide conversion in the synaptic neurones of the cerebrum of the rat in vitro. The ED_{50} amounted to 25.0 µmol so that of 5 sigma receptor ligands tested 1,3-di-o-tolylguanidine was second in regard to the strength of its effects (Bowen et al., 1988).

It was also possible to demonstrate in vitro that 1,3-di-o-tolylguanidine is bound to the sigma and phencyclidine (PCP) receptors in the frontal cortex membrane of the human brain. The concentration was 208 ± 85 nM for the sigma receptors and $54\,000 \pm 25\,000$ nM for the PCP receptors. 1,3-Di-o-tolylguanidine was thus bound specifically to the sigma receptors (Tam and Zhang, 1988).

7.11 Other effects

No information available.

8. Experience in humans

An older handbook reported 1,3-di-o-tolylguanidine as causing skin irritation and sensitization (Schwartz et al., 1957).

There was mention in a review that intoxication was unknown during commercial handling of 1,3-di-o-tolylguanidine (McCormick, 1971). The same information appeared in a safety data sheet (WTR, 1984).

Skin sensitization by 1,3-di-o-tolylguanidine could not be detected in patients with contact dermatitis due to rubber chemicals (van Dijk, 1968; Kantoh, 1985).

However, according to a further statement which was not verifiable, 1,3-di-o-tolylguanidine is said to have a sensitization effect (no further details; Schunk, 1984; Holmberg and Sjöström, 1977).

There were also reports of 3 female patients with dermatitis induced by Spandex, a synthetic polyurethane elastomer. 2 of these patients had in the past already had contact dermatitis that was attributed to rubber. Patch tests using 2% 1,3-di-o-tolylguanidine in Vaseline were among those conducted in all 3 of the female patients but in none of the cases was it possible to observe any positive reaction after 48 hours (van Dijk, 1968).

In order to identify the components of rubber products which have sensitization properties, a 1% concentration of 1,3-di-o-tolylguanidine was among the areas tested in the patch test. Observations made 72 hours after application revealed a positive reaction in none of the 13 patients who had suffered from rubber-induced contact dermatitis (Kantoh et al., 1985).

References

Bowen, W.D., Kirschner, B.N., Newman, A.H., Rice, K.C.
Sigma receptors negatively modulate agonist-stimulated phosphoinositide metabolism in rat brain
Eur. J. Pharmcol., 149, 399–400 (1988)

CIVO/TNO (Central Institute for Nutrition and Food Research, Zeist, Netherlands)
Toxicological examination of the compound WTR 28
Unpublished report no. R. 4850 (1975)

Contreras, P.C., Contreras, M.L., O'Donohue, T.L., Lair, C.C.
Biochemical and behavioral effects of sigma and PCP ligands
Synapse, 2, 240–243 (1988)

Dieke, S.H., Allen, G.S., Richter, C.P.
The acute toxicity of thioureas and related compounds to wild and domestic Norway rats
J. Pharmacol. Exp. Ther., 90, 260–270 (1947)

Dijk, van, E.,
Contact dermatitis due to Spandex
Acta Derm. Venereol., 48, 589–591 (1968)

Doehlemann, E.
Guanidin
In: Ullmanns Encyklopädie der technischen Chemie
4th ed., vol. 12, p. 411–420
Verlag Chemie, Weinheim (1976)

Doull, J., Plzak, V., Brois, S.J.
A survey of compounds for radiation protection
Armed Services Technical Information Agency, AD No. 277689 (1962)

Draize
J. Pharmacol., 82, 377–390 (1944)
Cited in: CIVO/TNO (1975)

Draize, Kelley
Drug Cosmet. Ind., 71, 36 (1952)
Cited in: CIVO/TNO (1975)

FDA (Food and Drug Administration)
Fed. Reg., 28 (119), 5582 (1963)
Cited in: CIVO/TNO (1975)

Hamburger, V., Hamilton, H.L.
A series of normal stages in the development of the chick embryo
J. Morphol., 88, 49–92 (1951)
Cited in: Korhonen et al. (1983 a)

Holmberg, B., Sjöström, B.
A toxicological survey of chemicals used in the Swedish rubber industry
Investigation report 1977: 19, pp. 11, 64
National Board of Occupational Safety and Health, Stockholm (1977)

Kantoh, H., Ishihara, M., Itoh, M., Hosono, K., Nishimura, M.
Allergens in rubber products
HIFU (Skin Research), 27 (3), 501–509 (1985)

Korhonen, A., Hemminki, K., Vainio, H.
Embryotoxic effects of acrolein, methacrylates, guanidines and resorcinol on three day chicken embryos
Acta Pharmacol. Toxicol., 52, 95–99 (1983a)

Korhonen, A., Hemminki, K., Vainio, H.
Toxicity of rubber chemicals towards three-day chicken embryos
Scand. J. Work Environ. Health, 9, 115–119 (1983b)

McCormick, W.E.
Environmental health control for the rubber industry
Rubber Chem. Technol., 44, 512–533 (1971)

NIOSH (National Institute for Occupational Safety and Health)
Guanidine, 1,3-di-o-tolyl-, RTECS-No. MF 1400000
Registry of Toxic Effects of Chemical Substances (1991)

Rannug, A., Rannug, U., Ramel, C.
Genotoxic effects of additives in synthetic elastomers with special consideration to the mechanism of action of thiurames and dithiocarbamates
Prog. Clin. Biol. Res., 141, Iss. Ind. Hazard Plast. Synth. Elastomers, 407–419 (1984)

Schunk, W.
Die Gefährdung von Arbeitern in der Gummiindustrie durch chemische Einsatzstoffe
Gummi, Fasern, Kunststoffe, 2, 58–64 (1984)

Schwartz, L., Tulipan, L., Birmingham, D.J.
Occupational diseases of the skin
3rd ed., Lea and Febinger, Philadelphia (1967)
Cited in: McCormick (1971)

Skofitsch, G.
Vom Hund zum Ei? Große und kleine Laboratoriumstiere
In: Alternativen zum Tierversuch
Ed. Fred Lembeck, Georg Thieme Verlag Stuttgart, New York, 71–79 (1988)

Smyth, H. F.
The toxicity of certain benzene derivatives and related compounds
J. Ind. Hyg., 13, 87–96 (1931)

Sumitomo Chemical Co. Ltd., Institute for Biological Science
Mutagenicity test of rubber chemicals
Unpublished report (1981)

Tam, S. W., Zhang, A. Z.
Sigma and PCP receptors in human frontal cortex membranes
Eur. J. Pharmacol., 154, 343–344 (1988)

Walker, J.M., Matsumoto, R.R., Bowen, W.D., Walker, F.O.
Dystonia in rats following microinjection of a sigma receptor ligand in the red nucleus
Neurology, 37 (Suppl. 1), 258–259 (1987)

Walker, J.M., Matsumoto, R.R., Bowen, W.D., Gans, D.L., Jones, K.D., Walker, F.O.
Evidence for a role of haloperidol-sensitive sigma-"opiate" receptors in the motor effects of antipsychotic drugs
Neurology, 38, 961–965 (1988)

WTR (International Working Group on the Toxicology of Rubber Additives)
Sicherheitsdatenblatt N,N'-Di-o-tolylguanidin (1984)

4-Nitrophenol

1. Summary and assessment

After intraperitoneal or intravenous injection into various species (rodents and poultry), 4-nitrophenol is rapidly distributed around the body and is excreted within 24 hours, essentially via the liver and kidneys in the urine. 4-Nitrophenol glucuronide and 4-nitrophenol sulphate occur as metabolites. Tests in dogs and rabbits reveal that up to 11 and 35%, respectively, is absorbed through the skin within 7 days, and metabolism and elimination are similar to that in other species.

4-Nitrophenol is of moderate toxicity after single oral or dermal exposures (LD_{50} rat oral 220 to 620 mg/kg body weight, dermal 1024 to 1300 mg/kg body weight). Symptoms observed after single doses include accelerated breathing, lying on the stomach or side, tonic-clonic spasms and episthotonus. Repeated oral administration of 70 mg/kg body weight or more reveals the liver and kidneys to be target organs. In addition, atrophy of the gonads and inhibition of motor activity are observed. 4-Nitrophenol does not cause methaemoglobin formation in cats.

4-Nitrophenol is mildly irritating to the skin of rabbits, while studies on the eye have shown severe irritancy.

In the maximization test in guinea-pigs, the compound has weak sensitizing potential.

Subchronic dermal application of 4-nitrophenol in doses of 50, 100 or 250 mg/kg body weight leads to dose-dependent macroscopic and microscopic damage to the skin (erythema formation, scaling, scurf and cracks) with corresponding histopathological changes (acanthosis, hypertrophy of the sebaceous glands, scurf formation, ulceration and swelling of the stratum spinosum). No signs of systemic intoxication are seen.

4-Nitrophenol shows no point-mutagenic activity in the majority of in vitro studies. Only in a few test systems positive mutagenic effects have been seen. Tests for point-mutagenic effects in vivo (in *Drosophila melanogaster*) are negative.

No increased tumour incidence is observed in male or female Syrian hamsters given 5 or 10 mg of a foundry effluent (which has contained 1800 µg 4-nitrophenol/g in addition to other phenols) by intratracheal administration, once a week for 18 weeks, followed by an observation period of 103 or 117 weeks. However, because of the relatively high acute toxicity and the resulting low doses used in this study, and bearing in mind the mixed exposure, this study is only of very limited value. p-Nitrophenol is not carcinogenic in mice after dermal application for 78 weeks.

According to the results of studies in mice, the compound has no embryotoxic or teratogenic effects. In addition, no effects on reproduction (fertility, gestation, live births, lactation) are observed in a 2-generation study in rats which involved dermal treatment for 100 or 120 days.

In vitro, the substance inhibits the cellular immune response (inhibition of the interleukin-2 system and lymphocyte proliferation following mitogenic stimulation).

4-Nitrophenol is cytotoxic to both animal and plant cells in vitro. The chemical inhibits conjugation with glucuronide and sulphate, although it is itself degraded by both of these metabolic steps. In contrast to many other phenols, the compound appears not to promote nitroso- or nitrite-formation from nitrogen dioxide.

Sensitization in man cannot be excluded. In medicine and in plant protection applications, 4-nitrophenol is used to detect the use of alkyl phosphates.

2. Name of substance

2.1	Usual name	4-Nitrophenol
2.2	IUPAC-name	4-Hydroxy-1-nitrobenzene
2.3	CAS-No.	100-02-7

3. Synonyms, common and trade names

4-Hydroxynitrobenzene
Niphen
4-Nitro-1-hydroxybenzene
para-Nitrophenol
p-Nitrophenol
PNP

4. Structural and molecular formulae

4.1 Structural formula

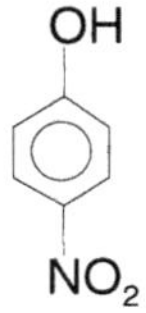

4.2 Molecular formula $C_6H_5NO_3$

5. Physical and chemical properties

5.1	Molecular mass, g/mol	139.11
5.2	Melting point, °C	114–116 (Weast, 1981/82) 112-114 (Hoechst, 1991)
5.3	Boiling point, °C	279 (decomposition) (Weast, 1981/82; Hoechst, 1991)
5.4	Vapour pressure, hPa	0.0032 (at 20 °C) 470 (at 70 °C) (Hoechst, 1991)
5.5	Density, g/cm³	1.479 (at 20 °C) (Weast, 1981/82)
5.6	Solubility in water	0.8 g in 100 g water (at 15 °C) (Thiem et al., 1979) 1.5 g in 100 g water (at 25 °C) (Hoechst, 1991) 29.1 g in 100 g water (at 90 °C) (Thiem et al., 1979)
5.7	Solubility in organic solvents	Dissolves easily in alcohol, ether, acetone (Thiem et al., 1979; Weast, 1981/82)
5.8	Solubility in fat	113-118 g/kg
5.9	pH-value	5 (Hoechst, 1991)
5.10	Conversion factor	1 ppm ≙ 5.68 mg/m³ 1 mg/m³ ≙ 0.18 ppm (at 25 °C and 1013 hPa)

6. Uses

Intermediate in the synthesis of photographic chemicals, dyestuffs, pesticides, fungicides and indicators; additive for petroleum and mineral oil; starting material for the manufacture of the analgesic 4-acetaminophenol (paracetamol; Thiem et al., 1979).

7. Experimental results

7.1 Toxicokinetics and metabolism

After oral administration (gavage) of 4-nitrophenol to rabbits (n = 3) 65% of the administered dose (0.2 g/kg body weight) were excreted as 4-nitrophenylglucuronide in the urine and 16% as 4-nitrosulphate within 24 hours. Less than 1% was excreted as free H-nitrophenol. Within 2 days an average of 87.4% of the administered dose was excreted as compounds with unchanged nitrogroup and 14.2% as amino compounds (total excretion 101.6%; the exceeding of 100% is probably due to the inexact colorimetric measurement method; Robinson et al., 1951).

In male Albany-Swiss mice (23 to 30 g, n = 6) 24 to 50% of the intraperitoneally injected dose (50 mg 4-nitrophenol/kg body weight) were excreted in the urine in a metabolized form within the first 5 hours. The main metabolites (95%) were glucuronic acid ester and sulphate ester. To a lesser extent the metabolites (1–2%) were identified as 4-nitrophenyl glucoside, which had not been found in mammals before (Gessner and Hamada, 1970). Minch et al. (1973) confirmed this finding in male Spague-Dawley rats.

Several in vivo studies in a number of species have shown that 4-nitrophenol is excreted in the urine after both glucuronidation and sulphation. Male and female Wistar rats (n = 4, age 60 days) were injected in the tail vein with 60 µmol (8.3 mg) 4-nitrophenol/kg body weight in 1 ml saline. The urine was collected on ice for 24 hours in 2-, 6- and 12-hour fractions. Bile was collected with a bile duct catheter. After various separation steps, the samples were studied with an absorption detector, and compared with the two metabolites, 4-nitrophenol glucuronide and 4-nitrophenol sulphate. It was shown that, in the urine, the adult rat excretes 33 to 34% of 4-nitrophenol as 4-nitrophenol sulphate and 37 to 39% as 4-nitrophenol glucuronide. No sex-specific differences were found. No details were provided of the analysis of the bile (Meerman et al., 1987).

In another study it was shown that 4-nitrophenol is not only metabolized in the liver, but also, to a not inconsiderable extent, in the kidneys. 4 male Sprague-Dawley rats (weight 335 to 400 g) had a catheter, to which an infusion pump was attached, inserted into the jugular vein. During the study, a solution containing 2.5% glucose and 0.45% sodium chloride was continually infused at a constant rate of 0.02 ml/min. Another catheter was inserted into the femoral artery in order to take blood samples, and both ureters and the main bile duct were similarly catheterized. The substrate, 4-nitro[2,6-^{14}C]-phenol (2 µmol/kg body weight), and the unlabelled conjugates, the glucuronide and the sulphate (0.3 µmol/kg body weight each), were simultaneously infused into the jugular vein for 90 minutes. A steady state were reached between 75 and 180 minutes after the start of the infusion, as shown by measurements of the concentration in the blood and the specific activity of the parent compound and the conjugates. After a 90-minute exposure period, 4 to 8 urine samples from each ureter were collected, each for periods of 15 minutes. Bile was collected in the same manner and subsequently analysed. Identification of ^{14}C-labelled 4-nitrophenol and its conjugates was achieved through separation by thin-layer chromatography with subsequent measurement of the activity by liquid scintillation. The recovery rates were 98, 97 and 95%. More than 98% of the infused ^{14}C-labelled 4-nitrophenol was recovered in the urine, and this was in the form of the sulphate and glucuronide conjugates. Renal metabolism accounted for a minimum of 20% of the endogenously formed conjugates with both conjugates of 4-nitrophenol formed to the same extent in the kidney (Tremaine et al., 1984).

The extra-hepatic conjugation of 4-nitrophenol in rats was demonstrated in an earlier in vivo study. Male Wistar rats (weight 200 to 350 g) had catheters placed in the left femoral vein and the left carotid artery. A catheter was also placed in the hepatic portal vein for comparative measurements. To collect samples of urine and bile, catheters were also inserted into the bladder and the bile duct. 4-Nitrophenol was then administered intravenously in physiological saline at 3 dose levels, 1.6, 4 and 8 mg/kg. The intra-arterial doses were 4 and 8 mg/kg, while those injected into the hepatic portal vein were 1.6 and 4 mg/kg. In all cases, blood samples were taken at 0, 1, 2, 4, 6, 8, 10, 20, 30, 60 and 90 minutes after the injection. The collection intervals for urine and bile were 0 to 4, 4 to 8 and 8 to 24 hours after the injection. Elimination of 4-nitrophenol from the plasma was very rapid. In all cases, the conjugates could be detected as early as 1 minute post-application: in the urine 37 to 41% was detected as the glucuronide and 30 to 61% as the sulphate, in the bile 6.2 to 11.1% was detected as the glucuronide. From the results obtained after administration into the femoral vein, it was concluded that conjugation was exclusively hepatic. The intra-arterial injection was designed to determine whether the lung was involved in 4-nitrophenol conjugation. The results did not differ from those obtained after intravenous application, and involvement of the lung could thus be excluded. Injection into the hepatic portal vein, however, produced different values from injection into the femoral vein, and extra-hepatic conjugation must therefore be assumed (Machida et al., 1982).

Sex-specific differences in glucuronidation were found in rats after intraperitoneal administration of ^{14}C-labelled 4-nitrophenol. Male and female Fischer-344 rats (age 12 to 15 weeks, weight 150 to 225 g) were given intraperitoneal injections of 0.05 µmol/kg ^{14}C-labelled 4-nitrophenol dissolved in 5 ml physiological saline/kg, and their urine was collected for 24 hours. In a second study protocol, rats of the same age and strain had catheters inserted into the jugular and carotid vessels as well as the bile duct and the urinary bladder. The rats were subsequently infused with ^{14}C-labelled 4-nitrophenol at a concentration of 2 µmol/minute/kg (activity 71.4 nCi/minute/kg) until a steady-state plasma concentration was reached. The glucuronide and sulphate conjugates were then infused at concentrations of 0.3 µmol/minute/kg for 90 minutes. 6 samples of urine and bile were collected at 10-minute intervals and analysed. In the meantime blood samples were taken. 24 hours after the intraperitoneal injection, ca. 51% of the administered dose was found in the urine of the male rats, and ca. 52% in the urine of the female rats. There were no sex-specific differences in the excretion of 4-nitrophenol sulphate, but the female rats excreted more 4-nitrophenol glucuronide in the urine than the males (ca. 36% and 14% of the concentration measured, respectively). There was also a sex-specific difference in the excretion of unchanged 4-nitrophenol. Male rats excreted ca. 41% of the chemical unchanged, while females excreted ca. 15%. The results of the second experiment with intravenous administration showed that, in vivo, the females excreted more 4-nitrophenol glucuronide in the urine than the males. This was mainly attributable to endogenous glucuronidation in the kidneys. The 4-nitrophenol glucuronide formed in the kidney and excreted in the urine was 2.7-fold greater in females than in males, and in females accounted for 22% of the total 4-nitrophenol glucuronide excreted. There were no sex-specific differences in the excretion of 4-nitrophenol glucuronide in the bile under these conditions. In total the proportion excreted in the bile was approximately 6% of the administered dose (Rush et al., 1983).

Extra-hepatic conjugation of 4-nitrophenol has also been demonstrated in poultry. Adult chickens (age 15 to 17 months, weight 3 to 4 kg), that had previously had catheters implanted in the wing vein and the leg artery, were given intravenous injections of 4-nitro[2,6-^{14}C]phenol (activity 34.8 mCi/mmol). Each ureteral opening in the cloaca was covered with a small plastic funnel in order to collect the urine. ^{14}C-Labelled 4-nitrophenol (250 nCi/minute) was continually infused into the cannulated wing vein. Inulin was infused into the wing vein with a separate needle at a rate of 2 mg/minute, in order to measure renal clearance. After an infusion period of 90 minutes, 10 urine samples were taken at intervals of 10 minutes. In the meantime, samples of arterial blood were also taken. In the second study, ^{14}C-labelled 4-nitrophenol with the same activity was infused into the leg vein, to study the first-pass effect in the kidney. From the leg vein, the substance could be excreted directly via the kidneys, thus leading to a greater increase in the urinary excretion of the parent compound if it was not metabolized or conjugated by the kidneys. On the other hand, there was an increased proportion of the corresponding conjugate in the urine, after conjugation did occur in the kidneys. About 93 to 94% of the infused material was recovered in the urine. Of this, 94.4% was identified as 4-nitrophenol sulphate, while the proportion of 4-nitrophenol glucuronide was 1.9% and that of unchanged 4-nitrophenol 1.7%. The specific activity of ^{14}C-labelled sulphate was greater in the urine than in the plasma, which indicated that the kidneys were involved in the conjugation. The conjugation in the kidneys was at least 38%. The clearance of 4-nitrophenol by the kidneys was found to be 9.81 ml/minute/kg/kidney, indicating rapid clearance of 4-nitrophenol from the plasma by conjugate formation and transport mechanisms (Diamond and Quebbemann, 1981).

A study of the excretion of 4-nitrophenol glucuronide in 2 dogs has been briefly reported. In this study, the plasma clearance of 4-nitrophenol glucuronide (33 and 39 ml/minute) was comparable with the inulin clearance (37±4 ml/minute in both animals). It was concluded from these results that, in dogs, glomerular filtration is the only renal excretion mechanism for 4-nitrophenol glucuronide (Hekman et al., 1986).

In another study, the absorption, distribution and excretion of 4-nitrophenol was determined in dogs and rabbits after intravenous injection or dermal application. 6 male New Zealand rabbits (weight 2.5 to 3.3 kg) and 6 male beagle dogs (25 months old) were used. 3 Rabbits were each given a single intravenous injection of ^{14}C-labelled 4-nitrophenol in the ear vein, at a dose of 357 µg (10 µCi per rabbit). The injection was carried out with the aid of a saline infusion. The remaining 3 rabbits had 4-nitrophenol applied dermally to their shaved backs. The area treated was 8.9 cm^2 and the dose applied was the same as that by intravenous administration (ca. 40.1 µg/cm^2). The treated area was then covered with a non-occlusive patch, which was changed after 24 hours. Similarly, 3 dogs were given single intravenous injections of 714 µg ^{14}C-4-nitrophenol (20 µCi per animal) with the aid of a saline infusion. The remaining 3 dogs had the same dose applied to the skin on their backs over an area of 17.8 cm^2 (40.1 µg/cm^2). The treated area was covered and handled in the same way. The half-life for the excretion phase was determined in the blood of the dogs after intravenous administration. Urine and faeces were collected at 24-hour intervals during the study, which lasted for 7 days. Aliquots of these were analysed by means of liquid scintillation after appropriate processing. At the end of the study, representative tissue and fluid samples were taken and the radioactivity of these was measured. Samples were studied from the liver, lungs, kidneys, spleen, heart, brain, adrenal glands, bladder, muscle, bones, skin, fat, thyroid gland, testes, bone marrow, blood and bile. The unabsorbed 4-nitrophenol on the skin was recovered by methanol extraction of the patches and the excised skin from the treated area. Measurements were then carried out by liquid scintillation. In both species, the substance was essentially excreted in the urine. The recovery rate for the radioactivity in the urine during the 7 days was 90% for the rabbits and 35% for the dogs. In both species, 96% of the radioactivity was excreted within the first 24 hours after the injection. The radioactivity in the faeces was 1% and <1%, respectively. The tissue samples taken from both species at autopsy contained no detectable ^{14}C-activity. The rate of excretion from the blood (half-life) was ca. 50 minutes in the dogs. After 4 hours, practically no radioactivity was measurable in the blood. After a single dermal application, an absorption rate of 35% of the applied dose was found during the 7-day study.

Of this, 78% was found in the urine within one day. The elimination was essentially complete after 4 days. The proportion of radioactivity in the faeces was less than 1%. The tissue samples contained practically no radioactivity. The unabsorbed ^{14}C-labelled 4-nitrophenol from the application site and the patch accounted for 53% of the applied dose. In dogs, the absorption rate after dermal application was 11%. An absorption rate of 3% per day was observed for the first 48 hours, and thereafter declined to less than 1% on day 5. The faecal excretion was negligible (<0.5% of the applied dose). Of the applied dose, 86% was not absorbed within 7 days. The tissue samples also contained no detectable radioactivity. Taking into consideration the fact that radiolabelled 4-nitrophenol was not detectable in the tissues of the dogs or the rabbits, the author concluded that 4-nitrophenol undergoes more-or-less complete metabolic clearance (Snodgrass, 1983).

In an in vitro study, the enzyme activity of UDP-glucuronyl transferase was measured after the addition of 5 mM 4-nitrophenol to liver homogenates from cows, sheep, pigs and rats. A decline in the activity of UDP-glucuronyl transferase was shown which was largest in the sheep, followed by the pig, the cow and the rat. Although the same mechanism was generally present in all species, there were marked differences in the order of magnitude of the activity. From this the authors concluded that extrapolation to other species should only be made with extreme caution (Smith et al., 1984). In the discussion in this paper, further data from the same institute is referred to, according to which dogs possess the highest transferase activity (Gregus et al., 1983).

In a further in vitro study in isolated liver cells from male Wistar rats, the distribution of the conjugation activity in the liver was studied after treatment with 4-nitrophenol. The isolated liver cells were separated into two different subpopulations (centrilobular and periportal hepatocytes) by means of the Percoll density-gradient method. The cells were then transferred to an appropriate medium containing 2.5 to 50 μmol 4-nitrophenol (to give 0.625×10^6 cells/ml) for 5 minutes. After the reaction was stopped, and appropriate processing had been carried out, the supernatant was analysed by HPLC. It was shown that glucuronidation occurred to a greater extent in the periportal liver cells than in the centrilobular hepatocytes (Michaelis-Menten constant higher by a factor of about 4). Comparable although less marked differences were found for sulphation (Araya et al., 1986).

In vitro a further metabolite of 4-nitrophenol was found, the 4-nitrocatechol, which was metabolized to the corresponding glucuronide or sulphate conjugate. In rat liver homogenate the formation of 4-nitrocatechol reached a maximum at pH 7.0 to 7.5; below pH 5 only a small amount of 4-nitrocatechol was formed (Chrastil and Wilson, 1975).

Livers from male Swiss mice (weight 20 to 30 g) were perfused for 45 minutes with 4-nitrophenol in a modified Krebs-Henseleit buffer at concentrations of 1, 8, 25 and 80 μmol. After a steady state was reached (15 minutes), 1 ml samples were taken at 10-minute intervals and the percentage proportions of the parent compound and metabolites were studied by means of HPLC. It was shown that, in the lower dose range (1 and 8 μmol), 4-nitrophenol was mainly metabolized by sulphation. With increasing concentration, sulphation declined and glucuronidation increased accordingly. The proportion of unmetabolized starting substance was increased in a similar manner (Sultatos and Minor, 1985).

The livers of anaesthetized male Wistar rats (weight 140 to 150 g) were perfused through the portal vein with 100 μmol 4-nitrophenol at a flow rate of ca. 4 ml/minute. To determine the output of 4-nitrophenol and glucose the hepatic vein was cannulated. Glucose was measured enzymatically and 4-nitrophenol photometrically. The perfusion was carried out for 1 hour. After 40 minutes, the perivascular nerves were stimulated for 5 minutes with a 2-pole platinum electrode. Measurements in the hepatic vein showed that ca. 50 μmol (corresponding to 50%) of the 4-nitrophenol was metabolized on passing through the liver, while the glucose level was only slightly reduced. During stimulation of the perivascular nerves the 4-nitrophenol level increased to about 70 μmol. The glucose level also clearly increased, while the flow rate was reduced. It was concluded by the authors that stimulation of the nerves had a direct effect on metabolism in the liver (Beuers et al., 1986).

The influence of hormones on the metabolism of 4-nitrophenol was studied in perfused livers of male Sprague-Dawley rats. The conjugation of 4-nitrophenol (initial concentration in the perfusate 60 μmol/l, corresponding to 8.3 mg/l) was less in hypophysectomized rats than in sham-operated control animals. Rates of glucuronidation were diminished by 55% (6.83 to 3.78 μmol/g/h), probably due to a reduced synthesis of UDP-glucuronic acid. Rates of sulphation were decreased by 50% (1.25 to 0.62 μmol/g/h) due to the reduced activity of the phenolsulpho transferase. Additional infusion of glucose was without effect in the sham-operated rats, but led to a slightly increased rate of conjugation in the hypophysectomized rats (Al-Turk and Reinke, 1983).

In vitro studies on liver microsomes from rats and rabbits have shown that 4-nitrophenol can be hydrolyzed to 4-nitrocatechol with the aid of the cytochrome-P450 enzyme complex (Reinke and Moyer, 1985; Koop, 1986).

In guinea-pig liver microsomes, in vitro studies have shown that in addition to phosphoglucuronic acid, uridine-diphosphogalacturonic acid is also involved in the metabolism of 4-nitrophenol. It was found that in addition to 4-nitrophenol glucuronide, 4-nitrogalacturonide was also formed (Vessey and Zakim, 1973).

The dependency of dermal adsorption on temperature was studied with the aid of a diffusion cell system in which the temperature could be regulated. The excised abdominal skin of male hairless mice (strain SKH-hr-1, age 60 to 80 days) was used as a membrane between the two compartments of the cell. In the first study the permeability was tested at temperatures of 10, 20, 30 and 37 °C in both chambers. The permeability of the skin to 4-nitrophenol increased with increasing temperature. In the second study, the temperature in the donor chamber was similarly varied, while the receptor chamber was maintained at 37 °C. The permeability to 4-nitrophenol was comparable with the results of the first study, from which the authors concluded that the permeability is independent of the donor temperature (Jetzer et al., 1988).

The permeation of different phenols was studied in vitro in isolated skin preparations of hairless mice (SHK-hr-1-strain, age 60 to 100 days) using a 2 chamber diffusion cell. Following an initial phase of 8.3 to 35.3 minutes 4-nitrophenol diffused well through the skin at pH 3.5. The permeability coefficient was 21×10^{-3} and 10×10^{-3} cm/h at a concentration of 5 and 0.5 g/l in the outer chamber, respectively. From measurements using skin preparations without stratum corneum it can be concluded that the permeability is mainly determined by the outer layer in contrast to the more hydrophobic phenols, for which the living tissue is a greater barrier (Huq et al., 1986).

In a further study in a diffusion cell the permeation of 4-nitrophenol was studied in excised skin from hairless mice, excised human skin and a silicone membrane. Contrary to expectations, very low permeability coefficients were found in all three studies. The octanol/water partition coefficient (log Pow) has a value of 1.96 and thus does not correlate with the low diffusion rate. From this the authors concluded that, in this case, the octanol/water partition coefficient is not appropriate for correlation with skin permeability. Finally they concluded that the principal pathway taken through the intact skin is of a lipophilic nature (Jetzer et al., 1986).

In vitro studies in isolated liver cells from Syrian hamsters and Fischer rats (males, 20 weeks old) showed that the glucuronidising activity in the hamster is considerably higher than that in the rat, whereas the sulphating activity in the rat is considerably higher than that in the hamster, indicating a species difference (Kokkinakis et al., 1987).

Other in vitro studies in isolated liver and mammary gland microsomes from lactating and non-lactating Sprague-Dawley rats have shown that 4-nitrophenol can also undergo glucuronidation in the mammary gland. The activity in the mammary gland was lower than that in the liver by a factor of about 3.4 in non-lactating rats. In lactating rats the difference was by a factor of 28 (Malejka-Giganti and Ryzewski, 1985).

In other in vitro studies in isolated hepatocytes it has been shown that pretreatment of hepatocytes from rats, hamsters and rabbits with ethanol interferes with the conjugation of 4-nitrophenol, and it is hydroxylated to 4-nitrocatechol (Reinke and Moyer, 1985; McCoy and Koop, 1988; Finely et al., 1986).

This effect has also been observed after pretreatment of isolated mouse hepatocytes with acetone (Banhegyi et al., 1988).

Two phenol sulfotransferases were detected in human liver homogenates, with the thermostable transferase having a considerably higher activity than the thermolabile transferase. In addition, this study showed that the phenol sulfotransferase does not consist of one uniform enzyme, but rather, of at least two isoenzymes (Campell et al., 1987).

In a series of further in vitro studies 4-nitrophenol served as substrate for the determination of the glucuronyltransferase activity (Jansen and Henderson, 1972; Chabra and Fouts, 1974; Grote et al., 1975; Moldéus et al., 1976; Reinke et al., 1981).

7.2 Acute and subacute toxicity

The available data on the acute toxicity of 4-nitrophenol are summarized in Table 1.

Table 1. Acute toxicity of 4-nitrophenol

Species	Sex [a]	Route and LD_{50} value (mg/kg body weight)	Reference
Rat		Oral: ca. 450 [b]	BASF, 1969
Rat		Oral: 570 (340–900) [c]	Vasilenko and Zvesdai, 1981
Rat	Female	Oral: 569	Hoechst, 1977 a
Rat	Male	Oral: 620 (450–850) [c]	Vernot et al., 1977
Rat	Male	Oral: 235 (within 24 hours)	Andrae et al., 1981 [d]
Rat	Male	Oral: 235 (within 14 days)	Andrae et al., 1981 [d]
Rat	Female	Oral: 235 (within 24 hours)	Andrae et al., 1981 [d]
Rat	Female	Oral: 20 (within 14 days)	Andrae et al., 1981 [d]
Rat	Female	Dermal: 1024 (984–1066) [c]	Hoechst, 1977 b
Rat	Male	Dermal: 1250 (within 24 hours/14 days)	Andrae et al., 1981 [d]
Rat	Male	Dermal: 1300 (within 24 hours/14 days)	Andrae et al., 1981 [d]
Mouse		Intraperitoneal: ca. 115 [b]	BASF, 1969
Mouse	Male	Intraperitoneal: 75[b]	Plzak and Doull, 1969
Mouse		Oral: 380 (220–650) [c]	Vasilenko and Zvesdai, 1981
Mouse	Male	Oral: 470 (320–690) [c]	Vernot et al., 1977

[a] Where specified
[b] Approximate LD_{50}
[c] 95% confidence limits
[d] In accordance with OECD guidelines

The administration of lethal doses of 4-nitrophenol to rats and mice led to motor disturbances, tonic spasms and accelerated breathing; death occurred due to cessation of breathing (BASF, 1969; Vasilenko and Zvesdai, 1981).

The oral LD_{50} in female Wistar rats (weight 87 to 120 g) was 569 mg/kg body weight. Signs of toxicity observed in the animals that died during the study included lying on the stomach or side, accelerated breathing, tonic-clonic spasms and episthotonus. Macroscopically, these animals had lungs which were grey-coloured, interspersed with dark red stains. Macroscopic examination of the surviving rats did not reveal any overt effects (Hoechst, 1977 a).

The LD_{50} values in male and female Sprague-Dawley rats (weight 165 to 170 g) after a single oral dose of 4-nitrophenol were 220 mg/kg body weight in the females and 235 mg/kg body weight in the males. All of the animals were examined macroscopically. The results of these examinations were, however, not reported (Andrae et al., 1981).

In cats (2 animals/dose) no haemoglobin formation was observed following a single oral dose of 100 to 500 mg 4-nitrophenol/kg body weight. After administration of 200 mg/kg 1 of 2 cats and after administration of 500 mg/kg 2 out of 2 cats died (BASF, 1969).

A dermal LD_{50} value of 1024 mg/kg body weight was determined in groups of 6 female Wistar rats (weight 150 to 198 g). The rats were treated with a single dose of 500 to 1500 mg 4-nitrophenol/kg body weight in sesame oil which was applied to their shaved, intact dorsal skin. The treated skin (ca. 30 cm^2) was covered occlusively for 24 hours. After the 24-hour exposure period, the dressing was removed and the skin was washed with a sponge and lukewarm water. The observation period was 14 days. Signs of toxicity observed inlcuded accelerated breathing and lying on the stomach. Death occurred 40 minutes to 18 hours after treatment. No specific adverse effects were observed on macroscopic examination (Hoechst, 1977 b).

In another study, the protocol for which corresponded to that described above, LD_{50} values after a single dermal application were 1300 mg/kg body weight for male rats and 1275 mg/kg body weight for female rats. No observations were made on signs of toxicity or macroscopic effects (Andrae et al., 1981).

In a study published in 1896, the lethal dose of 4-nitrophenol after subcutaneous administration was said to be 0.2 g/kg in rats and 0.2 to 0.3 g/kg in guinea-pigs (no further details; Binet, 1896).

An approximate LD_{50} of 75 mg/kg body weight was reported for mice that were given a single intraperitoneal injection of 4-nitrophenol. The observation period was 7 days (no further details; Plzak and Doull, 1969).

In an inhalation hazard test, inhalation of an atmosphere saturated with 4-nitrophenol at 20 °C for 8 hours was tolerated without signs of toxicity by a group of 12 rats. Post-mortems revealed no treatment-related effects (BASF, 1969).

In a preliminary study, non-pregnant CD-1 mice (61–71 days old, 10 animals/dose) received orally 0, 62.5, 125, 250, 500 or 1000 mg 4-nitrophenol/kg body weight/day in corn oil. From the cumulative mortality during the application period and the following observation period a LD_{50}-value of 626 mg/kg/day (95% confidence limit: 456–894 mg/kg/day) was calculated (see also section 7.8; Plasterer et al., 1985).

In a 28-day study, groups of 10 male and 10 female Sprague-Dawley rats (age 33 and 36 days, weight 100 to 112 g) were given 4-nitrophenol daily by stomach tube at doses of 70, 210 and 630 mg/kg body weight in 10 ml vehicle (0.8% aqueous hydroxypropyl methylcellulose gel, type E4M). At the end of the study, blood samples were taken from the retro-orbital venous plexus. For urine collection, the rats were placed in metabolic cages. Immediately prior to post-mortem, ophthalmic and hearing tests were carried out, and the teeth were inspected. The surviving animals were examined macroscopically. After treatment with 70 mg 4-nitrophenol/kg body weight/day for 28 days, pale livers were observed macroscopically in 7 of the rats, while histologically, 14 animals had fine droplets of fat deposited in their livers. These effects were thought to be treatment-related. All rats given 210 mg 4-nitrophenol/kg body weight/day developed symptoms such as lying on the stomach and decreased motor activity, from administration of the first dose onwards. Haematological tests revealed a trend towards a slight increase in the leukocyte count. 1 Male died on day 27 of the study in a comatose state. Macroscopically, 14 rats had pale livers, and microscopic examination revealed fatty degeneration. In addition, substance-related hydropic swelling of the liver was found in two cases. Repeated administration of the highest dose (630 mg 4-nitrophenol/kg body weight/day) led to behavioural effects comparable with those observed at the medium dose level, but considerably more marked. Haematological tests revealed a significantly increased leukocyte count in the females, and a corresponding trend was observed in the males. The clinical-chemical results in the males showed a slight increase in the activity of serum aminotransferase. During the course of the study, 4 males and 6 females died, but the cause of death could not be established. Histological examination of these animals revealed congestion of the liver and kidneys, and in addition, several cases of treatment-related nephrosis. The findings were significantly different from the controls. The occurrence of hydropic swelling of the liver, protein cylinders in the kidneys and atrophied testes, as well as the inhibition of spermiogenesis and follicle atresia were considered by the authors to the treatment-related. Bodyweight gain and feed and drink consumption appeared normal, and analyses of the faeces, of clinical-chemical and haematological parameters, and of urinary composition revealed no abnormalities. Examination of the eyes, ears and teeth gave no indications of treatment-related damage. In view of the low-dose effects, a no effect level could not be given (Andrae et al., 1981).

7.3 Skin and mucous membrane effects

In a skin irritation study, 500 mg of neat 4-nitrophenol was applied to the shaved flank of 6 albino Himalaya rabbits (weight 1.5 to 2 kg, sex unspecified) under occlusive cover for 24 hours. The effects were recorded immediately after removal of the dressing (24-hour value), and 48 and 72 hours after the beginning of the study. Evaluation was carried out with an irritation index from FDA guidelines (Federal Register, 38, No. 187, p. 27019, 1973). After application of the neat product, an irritation index of 2 was recorded, according to which the product is a mild skin irritant (Hoechst, 1977 c).

In a further tolerance test, which was carried out in accordance with OECD guideline no. 404, 6 rabbits (white Russian, 3 male and 3 female, weight at least 2.5 and 2.8 kg, respectively) had 0.5 g of the test substance applied to the shaved, intact dorsal skin under occlusive cover for 4 hours. After exposure, the dressings were removed, and the condition of the skin was evaluated after 30 and 60 minutes as well as after a further 24, 48, 72 and 96 hours, and after 7 and 14 days. There were no indications of skin irritation in this study (Andrae et al., 1981).

A 50% aqueous paste of technical grade 4-nitrophenol caused effects (examination after 24 hours) on the dorsal skin of rabbits after an exposure time of 1 minute. The findings resembled flea bites and were reversible after 8 days. Following a 5- or 15-minutes exposure slight erythema and bleeding was observed, and after 8 days scaling. An exposure time of 20 hours caused slight erythema, marked oedema, slight necrosis and bleeding. After 8 days a marked necrosis was seen. Slight erythema was seen, when the substance was applied to the ear for 20 hours (no information about the composition of the technical grade product; BASF, 1969).

After repeated dermal application of 50, 100 or 250 mg 4-nitrophenol/kg body weight in 0.5 ml ethanol daily (100 or 120 applications) to male and female Sprague-Dawley rats, dose-dependent damage to the skin was observed macroscopically and microscopically (further details of this study are reported in sections 7.5 and 7.8; Angerhofer, 1985).

The eye irritancy of 4-nitrophenol was studied in groups of 6 rabbits by instilling 100 mg of the neat product or 0.1 ml of a 10% dilution into the conjunctival sac of the left eye. The study was carried out in accordance with FDA guidelines (Federal Register, 38, No. 187, p. 27019, 1973). Evaluation was carried out with a magnifying glass, 1, 7, 24, 48 and 72 hours after instillation. After the examination at 24 hours, the eye was rinsed out with physiological saline. The examinations at 48 and 72 hours were made after instillation of 1 drop of fluorescein-sodium. The results were classified according to the Draize scale. Instillation of the neat product caused complete clouding of the cornea after only 1 hour in 5 of the 6 rabbits, and evaluation of the iris in these animals was not possible. In 1 rabbit, clouding of the lens was observed. In addition, slight redness and marked swelling of the conjunctiva were seen. These results indicate a severe irritant effect. Instillation of a 10% dilution of the product caused slight irritation (Hoechst, 1977 c).

Instillation of technical grade 4-nitrophenol as a powder into the eyes of rabbits caused slight oedema and clouding 1 hour after application. Corrosion of the eyeball and mucous membranes were seen. 24 hours after instillation severe oedema, and 8 days after instillation marked erythema, oedema and clouding as well as bleeding were observed (no information about the composition of the technical grade product; BASF, 1969).

In another eye irritation study carried out in accordance with OECD guideline 405, 100 mg or 0.1 ml of the test substance were instilled into the conjunctival sac of the left eye in 6 male rabbits (white Russian, weight 2.5 to 2.8 kg). The eyes were examined before instillation, and 1, 24 and 72 hours after instillation. 24 hours after instillation, the eyes were treated with fluorescein solution before being examined. Evaluation of the effects on the eye was carried out in accordance with the Draize scale. The results of this study gave no indications that 4-nitrophenol was an eye irritant (Andrae et al., 1981).

7.4 Sensitization

A maximization test (Magnusson and Kligman; OECD guideline 406) was carried out with 4-nitrophenol in male guinea-pigs (weight 250 to 300 g, age 21 to 28 days). The percentage of animals showing irritant reactions in the challenge phase in relation to the total number of animals tested was used for evaluation. The results of the study indicated that 4-nitrophenol had a weak sensitizing effect (5/20 animals; Andrae et al., 1981).

7.5 Subchronic and chronic toxicity

In the course of a reproductive toxicity study (the results of which are reported in Sect. 7.8), groups of 24 female and 12 male Sprague-Dawley rats (aged 4 weeks) were treated on the shaved dorsal skin, 5 times a week for 140 days (100 applications), as follows: group 1, vehicle (ethanol) 0.5 ml, group 2, control (physiological saline) 0.5 ml, group 3, 50 mg 4-nitrophenol/kg, group 4, 100 mg 4-nitrophenol/kg, group 5, 250 mg 4-nitrophenol/kg. The rats were observed daily for signs of toxicity, and body weight was recorded weekly. After gestation and lactation, the rats were dissected, the weights of the liver, kidneys, heart, gonads and brain were recorded, and sections from the skin and the gonads were taken and processed for histological examination. In an analogous study, groups of 26 females and 13 males from the F_1 generation were treated for 168 days (120 applications) and studied in the same way. All rats that were treated with 4-nitrophenol (F_0 and F_1 groups) showed dose-dependent damage to the skin, with erythema, sloughing, crusting and cracking. Histopathologically, chronic irritation, acanthosis, hypertrophy of the sebaceous glands, crust formation, ulceration and swelling of the stratum spinosum were found. The organ weights were comparable with those of the controls. Histopathologically, there were no specific effects on the internal organs. Substance related effects were only seen in the skin from a dose of 50 mg/kg (Angerhofer, 1985).

7.6 Genotoxicity

7.6.1 In vitro

The genotoxicity of 4-nitrophenol has been studied in numerous test systems. The study designs and results are presented in Table 2. In the majority of in vitro studies, 4-nitrophenol was not mutagenic. In a few test systems, however, positive mutagenic effects were seen.

7.6.2 In vivo

4-Nitrophenol was negative in two studies in *Drosophila melanogaster* (see Table 2; Vogel and Chandler, 1974; Zimmering et al., 1985).

7.7 Carcinogenicity

In a study carried out as part of the NTP program, groups of 60 male and 60 female Swiss-Webster mice received dermal applications of 0, 40, 80 or 160 mg 4-nitrophenol/kg (purity >97%, dissolved in acetone) on the interscapular region 3 times a week for 78 weeks. After 78 weeks the survival rates for the males were 29/60 (controls), 17/60 (40 mg/kg), 26/60 (80 mg/kg) and 24/60 (160 mg/kg), while in the females they were 35/60 (controls), 26/60 (40 mg/kg), 33/60 (80 mg/kg) and 27/60 (160 mg/kg). After 60 weeks, deaths were caused by generalized amyloidosis with secondary kidney failure, which occurred in both the treated animals and the controls. The body weights at the end of the study were comparable with the control groups in both sexes. Adverse effects due to 4-nitrophenol, in particular the occurrence of tumours, were not observed. There were thus no signs of 4-nitrophenol having carcinogenic activity under these study conditions. The Peer Review Panel agreed with this assessment, but the final report is not yet available (NTP, 1991).

In a study of the carcinogenic potential of foundry effluents, samples from the production process were collected and concentrated. Analysis of this sample revealed that it contained a number of phenols (phenol 1600 µg/g, pentachlorophenol <2 µg/g, 4-nitrophenol 1800 µg/g, 2-nitrophenol 2300 µg/g, 2,4-dimethylphenol 60 µg/g). In a dose-finding study, 20 Syrian hamsters (10 of each sex) had 20, 10, 5 or 2.5 mg of the sample in 0.2 ml distilled water instilled into the trachea once a week for 5 weeks. The maximum tolerated dose was 10 mg. In the subsequent chronic study, groups of 50 male and 50 female Syrian hamsters (7 weeks old) were treated with 10 or 5 mg of the emission sample in 0.2 ml distilled water which was instilled into the trachea once a week for 18 weeks. The hamsters were then kept until 98% of them had died. Two control groups of the same size were also studied. The first was treated with distilled water, and the second with a mixture of 3 mg benzo(a)pyrene and 3 mg iron trioxide, also in 0.2 ml distilled water. The study duration was 117 weeks for the males and 103 weeks for

Table 2. Genotoxicity of 4-nitrophenol

Indicator organism/ cell	Strain/ line	Type of test	Dose	Metabolizing system			Result		Dose-dependency of effect	Reference
				Type	with	with-out	with metabol.	without metabol.		
Salmonella typhimurium	TA 98	Reverse mutation	n.s.	S9-mix from rat liver	*	*	Negative	Negative	–	McCann et al., 1975
	TA 100				*	*	Negative	Negative	–	
	TA 1535				*	*	Negative	Negative	–	
	TA 1537				*	*	Negative	Negative	–	
Salmonella typhimurium	TA 98	Reverse mutation	n.s.	S9-mix from rat liver	*	*	Negative	Negative	–	Andrae et al., 1981
	TA 100				*	*	Negative	Negative	–	
	TA 1535				*	*	Negative	Negative	–	
	TA 1537				*	*	Negative	Negative	–	
	TA 1538				*	*	Negative	Negative	–	
Salmonella typhimurium	TA 98	Reverse mutation	0.8–2500 µl/ plate	S9-mix from rat liver	*	*	Negative	Negative	–	Hoechst, 1980
	TA 100				*	*	Negative	Negative	–	
	TA 1535				*	*	Negative	Negative	–	
	TA 1537				*	*	Negative	Negative	–	
	TA 1538				*	*	Negative	Negative	–	
Salmonella typhimurium	TA 98	Reverse mutation	100 µg/ plate	S9-mix from rat liver	*	*	Negative	Negative	–	Suzuki et al., 1983
	TA 100				–	*	–	Negative	–	
	TA 98			S9-mix from rat liver and Norharman	*	*	Negative	Negative	–	
Salmonella typhimurium	TA 98	Reverse mutation with pre-incu-	100–1000 µg/ plate	S9-mix from rat and hamster liver^	*	*	Negative	Negative	–	Haworth et al., 1983
	TA 100				*	*	Negative	Negative	–	
	TA 1535				*	*	Negative	Negative	–	
	TA 1537				*	*	Negative	Negative	–	

		bation								
Salmonella typhimurium	TA 98	Reverse mutation with pre-incubation	0.01–5 mg/plate	S9-mix from rat liver	*	*	Negative	Negative	–	Shimizu and Yano, 1986
	TA 100				*	*	Negative	Negative	–	
	TA 1535				*	*	Negative	Negative	–	
	TA 1537				*	*	Negative	Negative	–	
	TA 1538				*	*	Negative	Negative	–	
Salmonella typhimurium	TA 98	Reverse mutation with pre-incubation	100–5000 µg/plate	S9-mix from rat liver	*	*	Negative	Negative	–	Kawai et al., 1987
	TA 100				*	*	Negative	Negative	–	
Salmonella typhimurium	TA 98	Reverse mutation with pre-incubation	42–4173 µg/plate	S9-mix from hamster liver	*	–	Negative	–	–	Dellarco and Prival, 1989
	TA 100				*	–	Negative	–	–	
Salmonella typhimurium	TA 1535/pSK 1002	SOS	–745 µg/ml	S9-mix from rat liver	*	*	Negative	Negative	–	Nakamura et al., 1987
Salmonella typhimurium	TA 1538	Repair test	125–2000 µg/plate	–	–	–	Negative		–	Rashid and Mumma, 1986
	TA 1978			–	–	–	Negative			
Escherichia coli	K-12	Repair test	125–2000 µg/plate	–	–	–	Negative		–	Rashid and Mumma, 1986
	WP2									
Escherichia coli	WP2	Reverse mutation	0.8–2500 µl/plate	S9-mix from rat liver	*	*	Negative	Negative	–	Hoechst, 1980
Bacillus subtilis	H 17	Rec assay	0.05–0.5 mg/	–	–	–	Negative		–	Shimizu and Yano, 1986
	1145									

plate

Table 2. (continued)

Indicator organism/ cell	Strain/ line	Type of test	Dose	Metabolizing system			Result		Dose-dependency of effect	Reference
				Type	with	with-out	with metabol.	without metabol.		
Salmonella typhimurium	G 46	Host-mediated assay	75 mg/kg	–	–	*	–	Negative	–	Buselmaier et al., 1972
Proteus mirabilis	Wild type and repair-deficient strain	Difference in growth inhibition between strains	1–10 mg/ plate	–	–	*	–	Positive	n.s.	Adler et al., 1976
Hepatocytes (rat)	Isolated from rat liver	DNA-repair	0.03–3 mM	–	–	*	–	Negative	–	Andrae et al., 1981
Fibroblasts (human)	WI-38	DNA-repair	1–1000 μM	–	–	–	Positive		+	Poirier et al., 1975
Hepatocytes (rat)	Isolated from rat liver	UDS-test	100 nmol/ ml	–	–	–	Negative		–	Probst et al., 1981
Lymphoma cells (mouse)	L5178Y TK +/–	Forward mutation	58.4–782 μg/ml	S9-mix from rat liver	*	–	Negative	–	–	Amacher and Turner, 1982

Lymphoma cells (mouse)	L5178Y	Forward mutation	+S9 0.1–30 –S9 700–1500 µg/ml	S9-mix from rat liver	*	*	Negative	Negative	–	Oberly et al., 1984
Escherichia coli	Q13	DNA-binding	50 µM 500 µM	S9-mix from rat liver	* *	* *	Negative Positive	Negative Positive	+ +	Kubinski et al., 1981
Phages	Lambda	DNA-breaks	250 µM	–	–	–		Negative	–	Yamada et al., 1987
Saccharomyces cerevisiae	ade 2 trp 5	Mitotic gene conversion	21 mM	n.s.	n.s.	n.s.	Positive Positive		–	Fahrig, 1974
Drosophila melanogaster	n.s.	Recessive lethal mutations	n.s.	–	–	–	Negative		–	Vogel and Chandler, 1974
Drosophila melanogaster	Canton-S Wild type	Recessive lethal mutations	1000, 6000, 7500, 10000 ppm	In feed Injected	– –	– –	Negative Negative		– –	Zimmering et al., 1985

n.s. not specified – not performed or results not available
* performed + results available

the females. Histopathological examintion did not reveal any significant increase in the tumour incidence, particularly not in the respiratory tract, as would be expected. The authors concluded that because of the relatively high acute toxicity, the suspected carcinogen could only be administered at very low concentrations, which possibly explained the low tumour incidence (Moorman et al., 1985).

7.8 Reproductive toxicity

In a reproduction study with dermal pretreatment (as described in sect. 7.5) half the females in each group of rats were mated with males after the pretreatment period. Treatment of the females was continued during pregnancy and lactation. The F_1-generation was weaned 21 days after birth. The ovaries, uterus and samples of skin were taken from the females of the F_0-generation for histopathological examination. 26 females and 13 males per dose from the F_1-generation were chosen for continuation of the treatment and subsequent study. These rats were treated in the same manner as the F_0-generation, for 168 days (120 applications). At the end of this period, the animals of the F_1-generation were mated in the same way as the F_0-generation, and the offspring were kept until they were old enough to be weaned. During pregnancy and lactation, the rats were observed daily for signs of toxicity and behavioural effects, while the weights of the treated animals were recorded weekly. The observations during breeding and pregnancy included appearance and behaviour of the dams, litter size, individual foetal weights and viability on the day of birth, on days 4, 7 and 14 and at the time of weaning. The fertility index, gestation index, live births index and lactation index were calculated. The fertility index ranged from 58% (100 mg 4-nitrophenol dose group) to 87% (50 and 250 mg 4-nitrophenol dose groups). The fertility index of the control groups was also in this range. All pregnancies resulted in the birth of live offspring. Examination of the gonads did not reveal any treatment-related effects, but there were effects on the skin (see sect. 7.5). The survival rate in the F_1-generation from birth until weaning was nearly 100% in all dose groups. Body weight gain was normal in all of the F_1-rats. The fertility index in the F_1-generation ranged from 50% for the saline control group to 88% for the animals treated with 50 mg 4-nitrophenol. Again, all pregnancies led to the birth of live offspring. Apart from effects on the skin, no treatment-related effects were found. For the F_2-generation the survival rate for the period between birth and weaning was 96 to 100% in all groups. The animals, which were observed for a further 30 days after weaning, showed no abnormalities in behaviour or growth. Macroscopic examination revealed no adverse effects (Angerhofer, 1985).

In an embryotoxicity study, 50 pregnant mice (strain CD-1, 61 to 71 days old) were treated with 400 mg 4-nitrophenol/kg body weight by stomach tube from day 7 to day 15 of pregnancy. The body weight of the dams was recorded on day 7 and day 18 of pregnancy as well as on day 3 post partum. The foetuses were examined for macroscopically-visible changes on day 1 post partum. They were also weighed as whole litters on day 1 (12 to 24 hours after birth) and day 3 (48 hours after the first weighing). The dose used had been previously determined from the maximum tolerated dose. Treatment with 4-nitrophenol during the gestation period led to a maternal death rate of 19%, and in the surviving dams, to a reduction in body weight of 18.7%. Of the 36 pregnant dams, 29 survived, of which 28 produced live litters. Thus the reproductive index was 0.97. The numbers of surviving foetuses on day 1 and day 3 corresponded to the control values. The body weight of the litters were also comparable with the controls. Macroscopically, there were no detectable teratogenic effects (Plasterer et al., 1985).

In a study conducted to examine the effect of 4-nitrophenyl β-D-xyloside, an inhibitor of proteoglycan biosynthesis, on the growth of chick embryos a 4-nitrophenol/D-Xylose-mixture amongst others was injected (125 µl of a 6 g/l solution) into the white of chick eggs. The 4-nitrophenyl β-D-xyloside lead to abnormal growth and systemic oedema, while the 4-nitrophenol/D-xylose-mixture exhibited no effects (Oohira et al., 1981).

7.9 Effects on the immune system

The effect of 4-nitrophenol on the cellular immune response was studied in vitro in isolated spleen lymphocytes. The spleens from female Sprague-Dawley rats were processed, and the lymphocytes were separated by centrifugation and placed in an appropriate standard solution. 4-Nitrophenol was added to these cell cultures in defined doses. Proliferation of the cells was then initiated by the addition of the mitogen, concanavalin-A. Interleukin-2, an essential activator for the proliferation of T-lymphocytes during a cellular immune response, was then added to selected cultures. The effect of 4-nitrophenol on the interleukin-2 system was then studied with a specific mouse T-cell line (CTLL-2), which is absolutely dependent on interleukin-2 and the growth rate of which is proportional to the interleukin-2 concentration. 4-Nitrophenol clearly inhibited the mitogen-stimulated proliferation of lymphocytes, while the interleukin-2 system was also markedly inhibited. The authors concluded from this that 4-nitrophenol clearly suppresses the cellular immune response (Pruett and Chambers, 1988).

7.10 Neurotoxicity

No information available.

7.11 Other effects

4-Nitrophenol occurs as a metabolite of known compounds such as parathion, methylparathion and p-phenylmethane. The substance is predominantly detectable in the urine (Sultatos, 1986; Pena-Egido et al., 1988; Isobe, 1985).

4-Nitrophenol inhibits glucuronidation and sulphation, although the compound itself is excreted in the form of both of these conjugates (Tonda and Hirata, 1983).

Because 4-nitrophenol is a metabolite of specific pesticides, it may be viewed in the context of residue problems. Thus 4-nitrophenol was detectable in raw potatoes with a recovery rate of 82%. Even a cooking period of 30 minutes did not change the proportion of 4-nitrophenol in the potatoes (Ziegler et al., 1986).

In contrast to many other phenols, 4-nitrophenol did not appear to promote nitroso- or nitrite-formation from NO_2. The authors attributed this to the presence of the NO_2 group in 4-nitrophenol, which alters the typical properties (nucleophilicity, electron donation) of the aromatic ring (Cooney and Ross, 1987).

In vitro, 4-nitrophenol had a moderate cytotoxic effect on Chinese hamster ovary cells. The substance was added to cultures of the cells at concentrations of between 10 and 1000 µg/ml, 1 µCi ^{3}H-leucine and 0.5 µCi methyl-^{14}C-thymidine were added, and the mixture was then incubated for 20 hours at 39 °C. After incubation, protein synthesis and DNA synthesis were measured, and adenosine triphosphate consumption, cell numbers and cell viability were determined. The latter was measured with a haemocytometer after staining with trypan blue. Adenosine triphosphate was determined by means of fluorescence (luciferin-luciferase assay). The study showed that 4-nitrophenol had a marked effect on viability (reduction to 87% in relation to the control). However, more striking were effects on adenosine triphosphate and the synthesis rate of DNA and protein. The authors established that the capability to synthesize protein and DNA, taking ATP consumption into consideration, was a better indicator of cytotoxicity than the determination of cell viability (Garret and Lewtas, 1983).

In other studies in hamster fibroblasts and in *Vicia faba*, 4-nitrophenol was shown to be cytotoxic, causing disruption of protein and DNA synthesis (Packham et al., 1982; Amer and Ali, 1969).

8. Experience in humans

Dermal tests on 31 employees of a chemical factory revealed positive reactions to 4-nitrophenol in 4 males (Naniwa, 1979).

Cross reactions of dinitrochlorobenzene and other p-nitro compounds, including 4-nitrophenol, were not detectable in 27 patients (Eriksen, 1978).

Human kidneys derived from organ donors were perfused with ^{14}C-4-nitrophenol. At a perfusate concentration of 1.44 μmol/l the ^{14}C-activity in the 1 hour urine was the following: 21% unchanged 4-nitrophenol, 60% glucuronide and 10% sulphate. At a concentration of 0.72 μmol/l the sulphate conjugation comprised 31–56% of the radioactivity (Diamond et al., 1982).

In in vitro studies, 4-nitrophenol had a spermicidal effect in human ejaculate, destroying all sperm cells within 15 seconds at a concentration of 0.05 M (Mandal and Bhattacharyya, 1986).

In medicine, 4-nitrophenol is used as an indicator for intoxication with alkyl phosphates, particularly parathion and methylparathion (Michalke, 1984; Wehr, 1986; Stevens, 1986).

4-Nitrophenol also serves as an indicator of the incorporation of alkyl phosphates in the use of these pesticides in agriculture (detection as a metabolite in urine; Kummer and van Sittert, 1986).

References

Adler, B., Braun, R., Schöneich, J., Böhme, H.
Repair-defective mutants of Proteus mirabilis as a prescreening system for the detection of potential carcinogens
Biol. Zbl., 95, 463–469 (1976)

Al-Turk, W.A., Reinke, L.A.
Diminished conjugation of products of mixed-function oxidation in perfused livers from hypophysectomized rats
Pharmacology, 27, 74–84 (1983)

Amacher, D.E., Turner, G.N.
Mutagenic evaluation of carcinogens and non-carcinogens in the L5178Y TK assay utilizing postmitochondrial fraction (S9) from normal rat liver
Mutat. Res., 97, 49–65 (1982)

Amer, S.M., Ali, E.M.
Cytological effects of pesticides. IV. Mitotic effects of some phenols
Cytologica, 34, 533–540 (1969)

Andrae, U., Bieniek, D., Freitag, D., Göggelmann, W., Huber, W., Klein, W., Kotzias, D., Lahaniatis, E., Mansour, M., Parlar, H., Politzki, G., Rohleder, H., Rott, B., Scheunert, I., Spieser, H., Viswanathan, R.
Überprüfung der Durchführbarkeit von Prüfungsvorschriften und der Aussagekraft der Grundprüfung des E. Chem. G.
Gesellschaft für Strahlen- und Umweltforschung München mbH
On behalf of the Federal Office of the Environment (1981)

Angerhofer, R.A.
Effect of dermal applications of paranitrophenol on the reproductive functions of rats
US Army Environmental Hygiene Agency, report no. 75-51-0047-85 (1985)

Araya, H., Mizuma, T., Horie, T., Hayashi, M., Awazu, S.
Heterogeneous distribution of the conjugation activity of acetaminophen and p-nitrophenol in isolated rat liver cells
J. Pharmacobiodyn., 9, 218–222 (1986)

Back, K.C., Thomas, A.A., MacEwen, J.D.
Reclassification of materials listed as transportation health hazards
Natl. Tech. Inf. Serv. Rep. No. TSA-20-72-3, A 304–305 (1972)

Banhegyi, G., Garzo, T., Antoni, F., Mandl, J.
Accumulation of phenols and catechols in isolated mouse hepatocytes in starvation or after pretreatment with acetone
Biochem. Pharmacol., 37, 4157–4162 (1988)

BASF AG, Gewerbehygienisch-Pharmakologisches Institut
Gewerbetoxikologische Vorprüfung
Unpublished report (1969)

Beuers, U., Pogonka, T., Esterline, R., Ji, S., Jungermann, K.
Inhibition of para-nitrophenol extraction by stimulation of the hepatic nerves in the perfused rat liver
Toxicol. Lett., 34, 247–252 (1986)

Binet, P.
Toxicologie comparée des phénols
Revue médicale, 16, 449–465 (1896)

Buselmaier, W., Röhrborn, G., Propping, P.
Mutagenitäts-Untersuchungen mit Pestiziden im Host-mediated assay und mit dem Dominanten Letaltest an der Maus
Biol. Zbl., 91, 311–325 (1972)

Campbell, N.R.C., van Loon, J.A., Weinshilboum, R.M.
Human liver phenol sulfotransferase: assay conditions, biochemical properties and partial purification of isozymes of the thermostable form
Biochem. Pharmacol., 36, 1435–1446 (1987)

Chhabra, R.S., Fouts, J.R.
Sex differences in the metabolism of xenobiotics by extrahepatic tissue in rats
Drug Metab. Dispos., 2, 375–379 (1974)

Chrastil, J., Wilson, J.T.
4-Nitrocatechol production from p-nitrophenol by rat liver
J. Pharmacol. Exp. Ther., 193, 631–638 (1975)

Cooney, R.V., Ross, P.D.
N-Nitrosation and N-nitration of morpholine by nitrogen dioxide in aqueous solution: effect of vanillin and related phenols
J. Agric. Food Chem., 35, 789–793 (1987)

Dellarco, V.L., Prival, M.J.
Mutagenicity of nitro compounds in Salmonella typhimurium in the presence of flavin mononucleotide in preincubation assay
Environ. Mol. Mutagen., 13, 116–127 (1989)

Diamond, G.L., Quebbemann, A.J.
In vivo quantification of renal sulfate and glucuronide conjugation in the chicken
Drug Metab. Dispos., 9, 402–409 (1981)

Diamond, G.L., Anders, M.W., Tremaine, L.M., Quebbemann, A.J.
Salicylate enhancement of renal glucuronide conjugation and tubular excretory transfer of phenols
Drug Metab. Dispos., 10, 573–578 (1982)

Eriksen, K.
Cross allergy between paranitro compounds with special references to DNCB and chloramphenicol
Contact Dermatitis, 4, 29–32 (1978)

Fahrig, R.
Comparative mutagenicity studies with pesticides
In: Montesano, R., Tomatis, C., Davis, W. (eds.) IARC Sci. Publ., No. 10, p. 161–181
International Agency for Research on Cancer, Lyon (1974)

Finley, B.L., Ashley, P.J., Neptune, A.G., Yost, G.S.
Substrate-selective induction of rabbit hepatic UDP-glucuronyltransferase by ethanol and other xenobiotics
Biochem. Pharmacol., 35, 2875–2881 (1986)

Garret, N.E., Lewtas, J.
Cellular toxicity in Chinese hamster ovary cell culture. I. Analysis of cytotoxicity endpoints for twenty-nine priority pollutants
Environ. Res., 32, 455–465 (1983)

Gessner, T., Hamada, N.
Identification of p-nitrophenol glucoside as a urinary metabolite
J. Pharm. Sci., 59, 1528–1529 (1970)

Gregus, Z., Watkins, J.B., Thompson, T.N., Harvey, M.J., Rozman, K., Klaassen, C.D.
Hepatic phase I and phase II biotransformations in quail and trout: comparison to other species commonly used in toxicity testing
Toxicol. Appl. Pharmacol., 67, 430 (1983)

Grote, W., Schmold, A., Damman, H.G.
The metabolism of foreign compounds in rats after treatment with polychlorinated biphenyls (PCBs)
Biochem. Pharmacol., 24, 1121–1125 (1975)

Haworth, S., Lawlor, T., Mortelmans, K., Speck, W., Zeiger, E.
Salmonella mutagenicity test results for 250 chemicals
Environ. Mutagen., 5, Suppl. 1, 3–142 (1983)

Hekman, P., Russel, F.G.M., van Ginneken, C.A.M.
Renal transport of the glucuronides of paracetamol and p-nitrophenol in the dog
Drug Metab. Disp., 14, 370–371 (1986)

Hoechst AG, Pharma Forschung Toxikologie
Akute orale Toxizität von p-Nitrophenol an weiblichen SPF-Wistar-Ratten
Unpublished report no. 179/77 (1977 a)

Hoechst AG, Pharma Forschung Toxikologie
Akute dermale Toxizität von p-Nitrophenol an weiblichen SPF-Wistar-Ratten
Unpublished report no. 180/77 (1977 b)

Hoechst AG, Pharma Forschung Toxikologie
Haut- und Schleimhautverträglichkeit von p-Nitrophenol an Kaninchen
Unpublished report no. 181/77 (1977 c)

Hoechst AG, Pharma Forschung Toxikologie
Ames-Test 4-Nitrophenol
Unpublished report no. 538/80A (1980)

Hoechst AG
AIDA Grunddatensatz 4-Nitrophenol (1991)

Huq, A.S., Ho, N.F.H., Husari, N., Flynn, G.L., Jetzer, W.E., Condi, L., jr.
Permeation of water contaminative phenols through hairless mouse skin
Arch. Environ. Contam. Toxicol., 15, 557–566 (1986)

Isobe, M., Sone, T., Takabatake, E.
p-Nitrophenyl vinyl ether, a novel substrate for the assay of cytochrome P-450 dependent olefinic epoxidation in hepatic microsomes
J. Pharmacobiodyn., 8, 614–622 (1985)

Jansen, P.L.M., Henderson, P.T.
Influence of phenobarbital treatment on p-nitrophenol and bilirubin glucuronidation in Wistar rat, Gunn rat and cat
Biochem. Pharmacol., 21, 2457–2462 (1972)

Jetzer, W.E., Hou, S.Y.E., Huq, A.S., Duraiswamy, N., Ho, N.F.H., Flynn, G.L.
Temperature dependency of skin permeation of waterborne organic compounds
Pharm. Acta Helv., 63, 197–201 (1988)

Jetzer, W.E., Huq, A.S., Ho, N.F.H., Flynn, G.L., Duraiswamy, N., Condie, L.
Permeation of mouse skin and silicone rubber membranes by phenols: relationship to in vitro partitioning
J. Pharm. Sci., 75, 1098–1103 (1986)

Kawai, A., Goto, S., Matsumoto, Y., Matsushita, H.
Mutagenicity of aliphatic and aromatic nitro compounds
Jpn. J. Ind. Health, 29, 34–54 (1987)

Kokkinakis, D.M., Scarpelli, D.G., Subbarao, V., Hollenberg, P.F.
Species differences in the metabolism of N-nitroso(2-hydroxy propyl) (2-oxo-propyl)amine
Carcinogenesis, 8, 295–303 (1987)

Koop, D.R.
Hydroxylation of p-nitrophenol by rabbit ethanol inducible cytochrome P-450 isozyme 3a
Mol. Pharmacol., 29, 399–404 (1986)

Kubinski, H., Gutzke, G.E., Kubinski, Z.O.
DNA-cell-binding (DCB) assay for suspected carcinogens and mutagens
Mutat. Res., 89, 95–136 (1981)

Kummer, R., van Sittert, N.J.
Field studies on health effects from the application of two organophosphorus insecticide formulations by hand held ULV to cotton
Toxicol. Lett., 33, 7–24 (1986)

Machida, M., Morita, Y., Hayashi, M., Awazu, S.
Pharmacokinetic evidence for the occurrence of extrahepatic conjugative metabolism of p-nitrophenol in rats
Biochem. Pharmacol., 31, 787–791 (1982)

Malejka-Giganti, D., Ryzewski, C.N.
Microsomal metabolism of the carcinogen, N-2-fluorenyl-acetamide, by the mammary gland and liver of female rats. II. Glucuronidation of ring- and N-hydroxylated metabolites of N-2-fluorenyl-acetamide
Carcinogenesis, 6, 687–692 (1985)

Mandal, A., Bhattacharyya, A.K.
Human seminal antiliquefying agents – a potential approach towards vaginal contraception
Contraception, 33, 31–38 (1986)

McCann, J., Choi, E., Yamasaki, E., Ames, B.N.
Detection of carcinogens as mutagens in the Salmonella/microsome test: assay of 300 chemicals
Proc. Natl. Acad. Sci. USA, 72, 5135–5139 (1975)

McCoy, G.D., Koop, D.R.
Biochemical and immunochemical evidence for the induction of an ethanol-inducible cytochrome P-450 isozyme in male syrian golden hamsters
Biochem. Pharmacol., 37, 1563–1568 (1988)

Meerman, J.H.M., Nijland, C., Mulder, G.J.
Sex differences in sulfation and glucuronidation of phenol, 4-nitrophenol and N-hydroxy-2-acetylaminofluorene in the rat in vivo
Biochem. Pharmacol., 36, 2605–2608 (1987)

Michalke, P.
Determination of p-nitrophenol in serum and urine by enzymatic and non enzymatic conjugate hydrolysis and HPLC. Application after parathion intoxication
Z. Rechtsmed., 92, 95–100 (1984)

Minck, K., Schupp, R.R., Illing, H.P.A., Kahl, G.F., Netter, K.J.
Interrelationship between demethylation of p-nitroanisole and conjugation of p-nitrophenol in rat liver
Naunyn Schmiedebergs Arch. Pharmacol., 279, 347–360 (1973)

Moldéus, P., Vadi, H., Berggren, M.
Oxidative and conjunctive metabolism of p-nitroanisole and p-nitrophenol in isolated rat liver cells
Acta Pharmacol. Toxicol., 39, 17–32 (1976)

Moorman, W.J., Palmer, W.G., Garner, F.M., Mulligan, L.T.
Carcinogenic potential of foundry pyrolysis effluents
Am. Foundrymen's Society Trans., 93, 23–28 (1985)

Nakamura, S., Oda, Y., Shimada, T., Oki, I., Sugimoto, K.
SOS-inducing activity of chemical carcinogens and mutagens in *Salmonella typhimurium* TA 1535/pSK1002: examination with 151 chemicals
Mutat. Res., 192, 239–246 (1987)

Naniwa, S.
Industrial contact dermatitis due to nitro and amino derivatives. 1st report: mass examination of a factory
J. Dermatol., 6, 59–63 (1979)

NTP (National Toxicology Program)
Toxicology and carcinogenesis studies of p-nitrophenol in Swiss-Webster mice (dermal studies)
Tech. Report No. 417 (Draft) (1991)

Oberly, T.J., Bewsey, B.J., Probst, G.S.
An evaluation of the L5178YTK$^{+/-}$ mouse lymphoma forward mutation assay using 42 chemicals
Mutat. Res., 125, 291–306 (1984)

Oohira, A., Nogami, H., Nakanishi, Y.
Abnormal overgrowth of chick embryos treated with p-nitrophenol β-D-xyloside at early stages of development
J. Embryol. Exp. Morph., 61, 221–232 (1981)

Packham, E.D., Duxbury, C.L., Mayfield, C.I., Inniss, W.E., Kruuv, J., Thompson, J.E.
Quantitative analysis of pollutant induced lethal and sublethal damage in cultured mammalian cells
Bull. Environ. Contam. Toxicol., 29, 739–746 (1982)

Pena-Egido, M.J., Marino-Hernandez, E.L., Santos-Buelga, C., Rivas-Gonzalo, J.C.
Urinary excretion kinetics of p-nitrophenol following oral administration of parathion in the rabbit
Arch. Toxicol., 62, 351–354 (1988)

Plasterer, M.R., Bradshaw, W.S., Booth, G.M., Carter, M.W., Schuler, R.L., Hardin, B.D.
Developmental toxicity of nine selected compounds following prenatal exposure in the mouse
J. Toxicol. Environ. Health, 15, 25–38 (1985)

Plzak, V., Doull, J.
A further survey of compounds for radiation protection
Fed. Sci. Tech. Inf. Springfield VA 22151, AD 691490 (1969)

Poirier, M.C., DeCicco, B.T., Lieberman, M.W.
Nonspecific inhibition of DNA repair synthesis by tumor promotors in human diploid fibroblasts damaged with N-acetoxy-2-acetylaminofluorene
Cancer Res., 35, 1392–1397 (1975)

Probst, G.S., McMahon, R.E., Hill, L.E., Tompson, C.Z., Epp, J.K., Neal, S.B.
Chemically-induced unscheduled DNA synthesis in primary rat hepatocyte cultures: a comparison with bacterial mutagenicity using 218 compounds
Environ. Mutagen., 3, 11–32 (1981)

Pruett, S.B., Chambers, J.E.
Effects of paraoxon, p-nitrophenol, phenyl saligenin cyclic phosphate and phenol on the rat interleukin 2 system
Toxicol. Lett., 40, 11–20 (1988)

Rashid, K.A., Mumma, R.O.
Screening pesticides for their ability to damage bacterial DNA
J. Environ. Sci. Health, B21, 319–334 (1986)

Reinke, L.A., Bellinsky, S.A., Evans, R.K., Kauffman, F.C., Thurman, R.G.
Conjugation of p-nitrophenol in the perfused rat liver: the effect of substrate concentrations and carbohydrate reserves
J. Pharmacol. Exp. Ther., 217, 863–870 (1981)

Reinke, L.A., Moyer, M.J.
p-Nitrophenol hydroxylation – a microsomal oxidation which is highly inducible by ethanol
Drug Metab. Dispos., 13, 548–552 (1985)

Robinson, D., Smith, J.N., Williams, R.T.
Studies in detoxication. 39. Nitro compounds. (a) The metabolism of o-, m- and p-nitrophenols in the rabbit. (b) The glucuronides of the mononitrophenols and observations on the anomalous optical rotations of triacetyl β-o-nitrophenylglucuronide and its methylester
Biochem. J., 50, 221–227 (1951)

Rush, G.F., Newton, J.F., Hook, J.B.
Sex differences in the excretion of glucuronide conjugates: the role of intrarenal glucuronidation
J. Pharmacol. Exp. Ther., 227, 658–662 (1983)

Shimizu, M., Yano, E.
Mutagenicity of mono nitrobenzene derivatives in the Ames test and rec assay
Mutat. Res., 170, 11–22 (1986)

Smith, G.S., Watkins, J.B., Thompson, T.N., Rozman, K., Klaassen, C.D.
Oxidative and conjugative metabolism of xenobiotics by livers of cattle, sheep, swine and rats
J. Anim. Sci., 58, 386–395 (1984)

Snodgrass, H.L.
Dermal penetration and distribution of ^{14}C-labelled paranitrophenol
Study no. 75-51-0047-84
United States Army Environmental Hygiene Agency (1983)

Stevens, H.M.
The detection of some non-drug poisons in simulated stomach contents by diffusion into various colour reagents
J. Forensic Sci. Soc., 26, 137–145 (1986)

Sultatos, L.G.
The effects of phenobarbital pretreatment on the metabolism and acute toxicity of the pesticide parathion in the mouse
Toxicol. Appl. Pharmacol., 86, 105–111 (1986)

Sultatos, L.G., Minor, L.D.
Biotransformation of paraoxon and p-nitrophenol by isolated perfused mouse livers
Toxicology, 36, 159–169 (1985)

Suzuki, J., Koyama, T., Suzuki, S.
Mutagenicities of mononitrobenzene derivatives in the presence of norharman
Mutat. Res., 120, 105–110 (1983)

Thiem, K.W., Sewekow, B., Kiel, W., Handschuh, V., Freese, H., Schimpf, R., Vagt, H., Bunge, W.
Aromatische Nitroverbindungen
In: Ullmanns Encyklopädie der technischen Chemie, 4th ed., vol. 17, p. 409
Verlag Chemie, Weinheim (1979)

Tonda, K., Hirata, M.
Glucuronidation and sulfation of p-nitrophenol in isolated rat hepatocyte subpopulations. Effects of phenobarbital and 3-methylcholanthrene pretreatment
Chem. Biol. Interact., 47, 277–287 (1983)

Tremaine, L.M., Diamond, G.L., Quebbemann, A.J.
In vivo quantification of renal glucuronide and sulfate conjugation of 1-naphthol and p-nitrophenol in the rat
Biochem. Pharmacol., 33, 419–427 (1984)

Vasilenko, N.M., Zvesdai, V.I.
Possibilities of mathematical prediction of certain criteria of the toxicity of nitro and amino compounds of the benzene series (English translation)
Gig. Tr. Prof. Zabol., Iss. 8, 50–52 (1981)

Vernot, E.H., MacEwen, I.D., Haun, C.C., Kinkead, E.R.
Acute toxicity and skin corrosion data for some organic and inorganic compounds and aqueous solutions
Toxicol. Appl. Pharmacol., 42, 417–423 (1977)

Vessey, D.A., Zakim, D.
The identification of a unique p-nitrophenol conjugating enzyme in guinea pig liver microsomes
Biochem. Biophys. Acta, 315, 43–48 (1973)

Vogel, E., Chandler, J.
Mutagenicity testing of cyclamate and some pesticides in Drosophila melanogaster
Experientia, 30, 621–623 (1974)

Weast, R.C. (ed.)
CRC-Handbook of Chemistry and Physics
62nd ed., C-437, 1981/82
CRC-Press, Boca Raton, Florida

Wehr, K.
Zum Nachweis von E 605 nach mehrjähriger Erdgrabzeit
Z. Rechtsmed., 96, 57–66 (1986)

Yamada, K., Murakami, H., Yasumura, K., Shirahata, S., Shinohara, K., Omura, H.
Production of DNA-breaking substance after treatment of monophenols with sodium nitrite and then with dimethyl sulfoxide
Agric. Biol. Chem., 51, 247–248 (1987)

Ziegler, W., Hahn, M., Wallnöfer, P.R.
Untersuchungen zum Verhalten von Pflanzenschutzmittelrückständen in Wurzelgemüsearten unter den Bedingungen einer küchentechnischen Zubereitung: Glucose-konjugat von 3,4-Dichloranilin und von 4-Nitrophenol
Deutsche Lebensmittel-Rundschau, 11, 361–364 (1986)

Zimmering, S., Mason, J.M., Valencia, R., Woodruff, R.C.
Chemical mutagenesis testing in Drosophila. II. Results of 20 coded compounds tested for the National Toxicology Program
Environ. Mutagen., 7, 87–100 (1985)

2-Methyl-3-butyn-2-ol

1. Summary and assessment

According to the available animal data, 2-methyl-3-butyn-2-ol is of low acute toxicity on oral administration and inhalation (LD_{50} rat oral between 1420 und 1950 mg/kg body weight; LC_{50} rat >21 300 mg/m^3 air for 4 hours). On absorption, 2-methyl-3-butyn-2-ol has a narcotic effect.

An enriched atmosphere at 20 °C leads to irritation of the mucous membranes. In animals, depending on the duration of exposure, neat 2-methyl-3-butyn-2-ol is non-irritant or slightly irritating to the skin and severely irritating to the eyes.

There are no indications of sensitizing potential in the guinea-pig.

2-Methyl-3-butyn-2-ol shows no mutagenic acitivity in the Salmonella/microsome test.

On administration to rats by stomach tube for 4 weeks, the chemical causes sedation, leucocytosis, damage to the red blood cells, reduced growth, bleeding of the gastric mucosa and increased GPT-activity and cholesterol concentration in the plasma. The no effect level in rats is between 50 and 200 mg/kg body weight/day.

2. Name of substance

2.1	Usual name	2-Methyl-3-butyn-2-ol
2.2	IUPAC-name	2-Methylbutyn-3-ol-2
2.3	CAS-No.	115-19-5

3. Synonyms, common and trade names

3-Butin-2-ol, 2-methyl-
Dimethylacetylene-carbinol
Dimethylacetylenyl-carbinol
α,α-Dimethylpropargyl alcohol
1,1-Dimethylpropargyl alcohol
1,1-Dimethylpropynol
Ethynyldimethylcarbinol

2-Hydroxy-2-methyl-3-butyne
Isopentinol
3-Methyl-butyn-1-ol-3
2-Methylbutyn-3-ol-2
Methylbutynol
2-Methylbutin-3-ol-2

4. Structural and molecular formulae

4.1 Structural formula

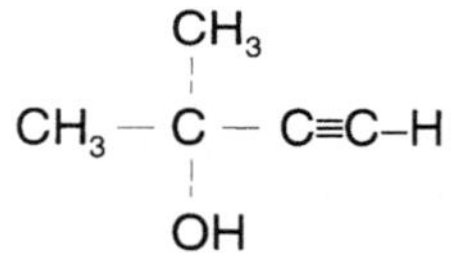

4.2 Molecular formula C_5H_8O

5. Physical and chemical properties

5.1	Molecular mass, g/mol	84.13
5.2	Melting point, °C	3 (solidification point) (BASF, 1986)
5.3	Boiling point, °C	102–105 (BASF, 1986)
5.4	Vapour pressure, hPa	20 (at 20 °C) (BASF, 1986)
5.5	Density, g/cm³	0.86 (at 20 °C) (BASF, 1986)
5.6	Solubility in water	Miscible (Windholz, 1983)
5.7	Solubility in organic solvents	Miscible with acetone, benzene, cellosolve (ethylene glycol ethyl ether), carbon tetrachloride, cyclohexanone, diethylene glycol, ethyl acetate, kerosene, methyl ethyl ketone, monoethanolamine, petroleum ether) (Windholz, 1983)

5.8 Solubility in fat	Soya bean oil (Windholz, 1983)
5.9 pH-value	Neutral (BASF, 1986)
5.10 Conversion factor	1 ppm ≙ 3.43 mg/m^3 1 mg/m^3 ≙ 0.29 ppm (at 25 °C and 1013 hPa)

6. Uses

Intermediate in the production of fragrances, cosmetics, vitamins, pesticides; stabilizer for halogenated hydrocarbons; used for protection against corrosion (BASF, 1987).

7. Experimental results

7.1 Toxicokinetics and metabolism
No information available.

7.2 Acute and subacute toxicity

After single exposures by the oral, intraperitoneal or subcutaneous routes, the LD_{50} values shown in Table 1 were determined.

In acute toxicological studies after oral or intraperitoneal administration, reeling, apathy, dyspnoea, ataxia, narcosis and lying on the side and stomach were observed in mice and rats (Brown et al., 1955; BASF, 1966). Autopsy revealed isolated cases of diarrhoea in the mice, but no effects in the rats (BASF, 1966).

Occlusive application of 0.2 ml neat 2-methyl-3-butyn-2-ol/kg body weight (172 mg/kg body weight) to the clipped dorsal skin of rabbits for 24 hours caused no local or systemic effects (BASF, 1966).

Repeated dermal application of 2-methyl-3-butyn-2-ol to mice, guinea-pigs and rabbits caused hyperaemia and dermatitis. Signs of systemic toxicity were not described (no further details; Balynina, 1987).

The LC_{50} value after a 2-hour exposure was given as 2000±150 mg 2-methyl-3-butyn-2-ol/m^3 air (582±44 ppm) for the mouse. Signs of toxicity included irritation of the upper airways, dyspnoea, loss of coordination and apathy (Balynina, 1987).

Table 1. Acute toxicity of 2-methyl-3-butyn-2-ol

Species	Number of animals/dose	Sex	Route of exposure	Observation period (days)	LD_{50} (95% confidence limits)	Reference
Mouse	n.s.	n.s.	p.o.	n.s.	500±30 mg/kg b.w.	Balynina, 1987
Mouse	50	n.s.	p.o. [a]	n.s.	1800 mg/kg b.w.	Keil et al., 1954
Mouse	n.s.	n.s.	i.p. [b]	7	ca. 1200 mg/kg b.w.	BASF, 1966
Mouse	n.s.	n.s.	i.p.	n.s.	3600 mg/kg b.w.	Windholz, 1983
Mouse	10	n.s.	s.c. [b]	24 hours	ca. 2340 mg/kg b.w.	Soehring et al., 1955
Mouse	n.s.	n.s.	Inhalation 2 hours	n.s.	2000±150 mg/m^3 air	Balynina, 1987
Rat	n.s.	n.s.	p.o. [b]	7	ca. 1420 mg/kg b.w.	BASF, 1966
Rat	10	n.s.	p.o.	n.s.	1950 mg/kg b.w. (1650–2300 mg/kg b.w.)	Brown et al., 1955
Rat	10	m/f	Inhalation 4 hours	14	>21 300 mg/m^3 air	BASF, 1988

n.s. not specified
[a] Dissolved in olive oil
[b] Aqueous solution

In an inhalation-hazard test, an enriched atmosphere at 20 °C (no details of concentration) was inhaled by groups of 6 or 12 rats for 30 minutes, 1 hour or 4 hours. The enriched atmosphere caused irritation of the mucous membranes and narcosis (no further details). A 30-minute exposure was tolerated, but 2/12 rats died on exposure for 1 hour and all 6 rats died on exposure for 4 hours. Macroscopic examination at autopsy revealed hydrothorax (BASF, 1966).

5 Male and 5 female rats were exposed to an average 21.28 (18.8 to 22.7) mg 2-methyl-3-butyn-2-ol/l air (corresponding to 6193 ppm) for 4 hours in a dynamic whole-body inhalation chamber. During exposure, accelerated and intermittent breathing occurred, together with paleness, narcosis and lying on the side. After exposure, visible signs of toxicity included lying on the side or stomach, narcosis, accelerated or irregular breathing, general malaise, apathy, reeling or unsure gait, ruffled fur, red crusts around the edges of the eyes and urine-smeared anogenital regions. Weight gain was reduced. From day 9 of the observation period, no further signs of toxicity were observed. Pathological examination after 14 days of observation gave no indication of substance-related effects. Thus the 4-hour LC_{50} value was >21280 mg/m^3 (BASF, 1988).

Rats (BOR:WISW/SPF Cpb) were given 50, 200 or 800 mg 2-methyl-3-butyn-2-ol/kg body weight/day by stomach tube for 4 weeks. 10 Rats of each sex were used per dose. 10 Further animals of each sex served as controls. Marked signs of toxicity were seen in the high-dose group, so these animals were not treated on day 3 of the study and from day 4 they were given 600 mg/kg body weight/day. Water was used as a solvent, and the volume administered was 5 ml/kg body weight. The highest dose level caused marked sedation and narcosis, hypothermia, reduced feed consumption (−13%), reduced body weight gain and increased mortality (5 females and 3 males). A significant ($p \leq 0.05$) reduction in the leukocyte count and a reduced haemoglobin concentration were found in the females at the highest dose level, on analysis of the blood of 5 males and 5 females from each group at the end of the study. The thrombocyte count, haemocrit value, reticulocyte count, median cell volume (MCV), median cellular haemoglobin content (MCH) and median cellular haemoglobin concentration (MCHC) were unaffected. The differential blood count revealed a significant ($p \leq 0.01$) increase in the percentage rise in neutrophils with segmented nuclei, and a corresponding decrease in the lymphocytes, in the males at 200 mg/kg and above and in both sexes at the highest dose level. In all the males and females studied, there was a dose-dependent increase in polychromasia from 200 mg/kg. In addition, in the females given 200 mg/kg or more and in the males given 800/600 mg/kg body weight, a dose-dependent increase in erythrocyte anisocytosis was seen. Moreover, many anulocytes were observed at the high dose level (no further details). Clinical-chemical studies revealed a significant ($p \leq 0.05$) increase in cholesterol and in glutamate-pyruvate transaminase activity in the plasma in both sexes. In addition, the concentrations of bilirubin and total protein in the plasma were significantly increased ($p \leq 0.05$) in the females. The activities of alkaline phosphatase and glutamate oxaloacetate transaminase were within the range of the controls, as were the levels of creatinine, urea, glucose, sodium, potassium, calcium, chloride and inorganic phosphate. Post-mortems on the rats that died prematurely (high-dose group) revealed hyperaemia of the lungs, focal reddening of the mucous membrane of the fundus of the stomach and blood-red colouration of the small intestine contents. Histologically, the mucosal reddening corresponded to hyperaemia, focal bleeding in the mucous membrane and, in some cases, focal erosive or ulcerative gastritis. No morphological effects were observed either macroscopically or microscopically in the animals killed at the end of the study; histological studies were made of the brain, heart, bone and bone marrow (femur, sternum), and of any other organs that appeared abnormal on gross examination. The organ weights were within the control range. Under these study conditions, the no effect level was between 50 and 200 mg/kg body weight/day (Bayer, 1984).

7.3 Skin and mucous membrane effects

The acute skin irritancy was studied in a patch test on the clipped dorsal skin of white rabbits, in which a cotton patch (ca. 2.5×2.5 cm) soaked with the neat substance (no further details) was in contact with the skin for 1, 5 or 15 minutes or for 20 hours. After the 1-, 5- and 15-minute applications, the treated skin was first washed with neat polyethylene glycol 400 and then with a 50% aqueous solution of polyethylene glycol 400. After exposure for 20 hours, the skin was left untreated. No changes were evident 24 hours after the 1- and 5-minute exposures. The 15-minute exposure resulted in marginal reddening and the 20-hour exposure in slight reddening, but these effects had cleared up completely after 8 days (BASF, 1966).

To study the acute eye irritancy, 50 µl neat 2-methyl-3-butyn-2-ol was instilled into the conjunctival sac and carefully distributed over the surface of the rabbit's eye. 1 Hour after instillation, the test substance had caused marked redness and severe oedema formation as well as marked clouding of the cornea. After 24 hours, bleeding was observed in the conjunctiva. 8 Days after instillation, severe redness and slight clouding of the cornea were still present, as well as a staphyloma. The sodium chloride instilled into the other eye as a control had no effect (BASF, 1966).

According to another report, the substance is severely irritating to the eye, causing swelling, purulent kerato-conjunctivitis and clouding of the cornea (no further details; Balynina, 1987).

7.4 Sensitization

2-Methyl-3-butyn-2-ol was applied to the clipped dorsal skin of 10 guinea-pigs as 4 doses of 0.1 ml over 10 days. In addition, 0.2 ml Freund's adjuvant was administered intradermally next to the skin application site at the same time as the third dose was applied (modified split-adjuvant test). A challenge with 0.1 ml of the neat substance on the clipped flank after 14 days gave no indications of a sensitizing effect (Rao et al., 1981).

7.5 Subchronic and chronic toxicity

No information available.

7.6 Genotoxicity

7.6.1 In vitro

The mutagenic potential of 2-methyl-3-butyn-2-ol was tested in prokaryotes (*Salmonella typhimurium* strains TA 98, TA 100, TA 1535, TA 1537 and TA 1538). Concentrations of up to 2500 µg/plate were used. No toxic effect was observed up to the highest concentration employed. The studies were carried out with and without metabolic activation (S9-mix from Aroclor 1254-induced rat liver). There were no indications of mutagenic activity (BASF, 1980).

7.6.2 In vivo

No information available.

7.7 Carcinogenicity

No information available.

7.8 Reproductive toxicity

No information available.

7.9 Effects on the immune system

No information available.

7.10 Neurotoxicity

No information available.

7.11 Other effects

The threshold value for 50% haemolysis in vitro was 0.67 mol/l (equivalent to 56.4 mg/ml) in human blood and 0.6 mol/l (equivalent to 50.5 mg/ml) in canine blood (Soehring et al., 1955).

2-Methyl-3-butyn-2-ol has mild CNS-depressant properties in animals (Brown et al., 1955; Soehring et al., 1955).

The subcutaneous ED_{50} for a 5 minute lying on the side, was 780 mg/kg body weight after administration of 2-methyl-3-butyn-2-ol, and 100 mg/kg after administration of phenobarbital in the mouse (Soehring et al., 1955).

Subcutaneous injection of 1200 mg 2-methyl-3-butyn-2-ol/kg body weight led to a mean sleeping time of 275 minutes in mice after an average of 5.8 minutes, whilst a dose of 115 mg phenobarbital/kg caused mice to fall asleep for 130 minutes after a delay of 38 minutes (Soehring et al., 1955).

Injection of 700 mg/kg body weight s.c. led to a 1.9-fold increase in the threshold for convulsions triggered by cardiazol (Soehring et al., 1955).

2-Methyl-3-butyn-2-ol had an anti-convulsant effect in the rat. The oral ED_{50} for spasms induced by electric shocks was determined as 140 mg/kg body weight; for phenobarbital it was 19 mg/kg (Brown et al., 1955).

8. Experience in humans

No information available.

References

Balynina, E.C.
Toxizität von Dimethylethinylcarbinol
Gig. Tr. Prof. Zabol., 0 (4), 55 (1987)

BASF AG, Gewerbehygienisch-Pharmakologisches Institut
Methylbutinol – Ergebnis der Gewerbetoxikologischen Vorprüfung
Unpublished report XVI/229 (1966)

BASF AG, Gewerbehygiene und Toxikologie
Bericht über die Prüfung von 2-Methyl-butin-3-ol-2 im Ames-Test
Unpublished report (1980)

BASF AG
2-Methylbutin-3-ol-2 – DIN-Sicherheitsdatenblatt (1986)

BASF AG
Grunddatensatz für Großstoffe, 2-Methylbutin-3-ol-2 (1987)

BASF AG, Abteilung Toxikologie
Bericht über die Prüfung der akuten Inhalationstoxizität LC_{50} von 2-Methylbutin-3-ol-2 als Dampf an Ratten, Exposition 4 Stunden
Unpublished report (1988)

Bayer AG, Institut für Toxikologie
2-Methylbutin-3-ol-2
Bericht über subakute toxikologische Untersuchungen an Ratten (Versuch mit Magensondenapplikation über 4 Wochen)
Unpublished report (1984)

Brown, B., Schaffarzick, R.W., Dreisbach, R.H.
Anticonvulsant properties of certain secondary and tertiary alcohols
J. Pharmacol. Exp. Ther., 115, 230–239 (1955)

Keil, W., Muschaweck, R., Rademacher, E.
Die sedative und hypnotische Wirkung des 1-Äthinyl-cyclohexyl-allophanat-(1)
Arzneimittelforschung, 4 (8), 447–479 (1954)

Rao, K.S., Betso, J.E., Olson, K.J.
A collection of guinea pig sensitization test results – grouped by chemical class
Drug Chem. Toxicol., 4 (4), 331–351 (1981)

Soehring, K., Frey, H.H., Endres, G.
Der Zusammenhang zwischen Konstitution und Wirkung tertiärer Alkohole
Arzneimittelforschung, 5 (4), 161–165 (1955)

Windholz, M. (ed.)
The Merck Index
10th ed., p. 866
Merck & Co., Inc., Rahway (1983)

N-Carboxyanthranilic acid anhydride

1. Summary and assessment

N-Carboxyanthranilic acid anhydride is slightly toxic following a single oral or dermal dose (oral LD_{50} rat: >2500 mg/kg body weight; dermal LD_{50} rat: >500 mg/kg body weight; LC_{50} rat, 4 hours: >5300 mg/m^3). In the inhalation risk test in rats there are no instances of death on exposure for 8 hours.

N-Carboxyanthranilic acid anhydride is non-irritant to the skin of rabbits, but is irritant to the eye.

In rats, inhalation of 11.6 and 71 mg/m^3 air for 4 hours/day, 5 days/week for 4 weeks produces no substance-related changes in organ weights, apart from an increase in absolute and relative lung weight and an increase in relative liver weight in the female animals in the higher dose group. There are no results reported on histopathological examinations. The hematological and biochemical parameters in the two test groups are not different to those of the controls. 11.6 mg/m^3 are tolerated without symptoms.

N-Carboxyanthranilic acid anhydride has a sensitizing effect in guinea-pigs.

N-Carboxyanthranilic acid anhydride is neither mutagenic in *Salmonella typhimurium,* with or without metabolic activation, nor in the micronucleus test after oral administration to mice.

When ingested orally, N-carboxyanthranilic acid anhydride can apparently lead to impairment of fertility in female mice. The existing study does not however permit any conclusive evaluation.

N-Carboxyanthranilic acid anhydride is not irritant to human skin. No sensitizing effects have been observed in the respiratory tract or in the skin of workers who can be exposed to N-carboxyanthranilic acid anhydride during production processes.

2. Name of substance

2.1	Usual name	N-Carboxyanthranilic acid anhydride
2.2	IUPAC-name	2H-3,1-Benzoxazine-2,4-(1H)-dione
2.3	CAS-No.	118-48-9

3. Synonyms, common and trade names

N-Carboxyanthranilic acid cyclic anhydride
2-(Carboxyamino)benzoic acid cyclic anhydride
2,4-Dioxo-1,2-dihydro-4H-3,1-benzoxazine
Isatoic acid anhydride
Isatoic anhydride
Isatosäureanhydrid
N-Carboxyanthranilic anhydride
N-Carboxyanthranilsäureanhydrid

4. Structural and molecular formulae

4.1 Structural formula

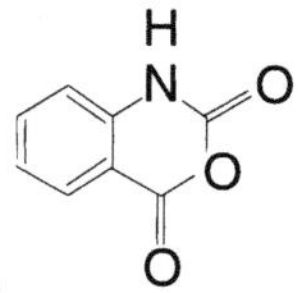

4.2 Molecular formula $C_8H_5NO_3$

5. Physical and chemical properties

5.1	Molecular mass, g/mol	163.13
5.2	Melting point, °C	230–232 (BASF, 1987)
5.3	Boiling point, °C	decomposition from 350 (BASF, 1987)
5.4	Vapour pressure, hPa	no information available
5.5	Density, g/cm^3	approx. 600 kg/m^3 (bulk density) (BASF, 1987)
5.6	Solubility in water	approx. 300 mg/l (at 20 °C) (BASF, 1987)

5.7 Solubility in organic solvents	slightly soluble in ethanol (Bayer, 1970) soluble in dimethylformamide (BASF, 1988)
5.8 Solubility in fat	no information available
5.9 pH-value	no information available
5.10 Conversion factor	1 ppm ≙ 6.66 mg/m³ 1 mg/m³ ≙ 0.15 ppm (at 25 °C and 1013 hPa)

6. Uses

Intermediate for dyes, perfumes, flavours, drugs and crop protection chemicals (BASF, 1987).

7. Experimental results

7.1 Toxicokinetics and metabolism

No information available.

7.2 Acute and subacute toxicity

The following LD_{50} values were found following single oral, intraperitoneal and dermal administration (see also table 1):

Mouse	oral	>1000 mg/kg body weight
Mouse	i.p.	between >1000 and 1250 mg/kg body weight
Rat	oral	>2500 mg/kg body weight
Rat	dermal	>500 mg/kg body weight
Rat	i.p.	810 mg/kg body weight
Guinea-pig	oral	approx. 500 mg/kg body weight
Cat	oral	>500 mg/kg body weight
Hen	oral	>2500 mg/kg body weight

Impairment in general health (no further details), dyspnoea and, in some cases, clonical spasms appeared as intoxication symptoms in mice, rats, guinea-pigs and cats following oral or intraperitoneal administration. Following oral administration the symptoms appeared at the earliest after 60 minutes, and receded in sur-

Table 1. Acute toxicity of N-Carboxyanthranilic acid anhydride

Animal species	Number of animals/ dose	Sex	Route of adminis- tration	Post-treatment observation period (days)	LD_{50} (mg/kg b.w.)	LD_{50} (95% confi- dence limit) (mg/kg b.w.)	References
Mouse	15	Male	Oral	14	>1000	>1000	Bayer, 1970
Mouse	15	Male	i. p.	14	500	>1000	Bayer, 1970
Mouse	n.d.	n. d.	i. p.	n.d.	n.d.	1250	BASF, 1971
Mouse	20	Male	Inhal.[a]	14	>378 mg/m^3	>378 mg/m^3	Bayer, 1970
Rat	15	Male	Oral	14	>2500	>2500	Bayer, 1970
Rat	n.d.	n. d.	Oral	n.d.	n.d.	>5000	BASF, 1971
Rat	15	Male	i. p.	14	500	810	Bayer, 1970
Rat	5	Male	Dermal	7	n.d.	>500	Bayer, 1970
Rat	20	Male	Inhal.[a]	14	>255 mg/m^3	>255 mg/m^3	Bayer, 1970
Rat	20	Female	Inhal.[a]	14	>348 mg/m^3	>348 mg/m^3	Bayer, 1970
Guinea-pig	5	Male	Oral	14	250	ca, 500	Bayer, 1970
Cat	2	Male	Oral	14	>500	>500	Bayer, 1970
Hen	5	n. d.	Oral	14	>2500	>2500	Bayer, 1970

b. w. = body weight

n. d. = no details

[a] 4-hour-exposure

viving animals after a maximum of 5 days. Death occurred within 3 days. Following intraperitoneal administration the first symptoms of intoxication were observed after 5 minutes, receding in the surviving animals within 4 days. Mortality occurred within 1 to 3 days. Following dermal administration no symptoms of intoxication could be observed in the 7-day study period (Bayer, 1970).

In the inhalation-risk test in rats there were no instances of death on exposure to an atmosphere enriched with N-carboxyanthranilic acid anhydride dust for 8 hours (BASF, 1971).

In a limit-test according to OECD-guideline 403 and LC_{50} of >5.3 mg/l (>5300 mg/m³) was determined in Wistar rats after head/nose-only exposure for 4 hours, particle size analysis showing that 81% of the dust-aerosol fraction reached the alveolar region. Only in some animals was irregular or rapid breathing and reddish fluid at the noses observed as clinical symptoms during the exposure time. During the 14-day observation period no substance related pathological effects were seen. Body weight gain did not differ from the controls (BASF, 1990 a).

The following LC_{50} values were determined following dynamic inhalative exposure to N-carboxyanthranilic acid anhydride, sprayed in a mixture with DMSO and Lutrol (1:1), for 4 hours: mouse, male, >378 mg/m³ air; rat, male, >255 mg/m³ air; rat, female, >348 mg/m³ air. Air analyses were conducted in the inhalation chamber in order to determine the concentration. In rats and mice, 4 hours exposure to 255, 348 and 378 mg/m³ resulted in temporary impairment in general condition after exposure ceased. Deaths have not been observed (no further details; Bayer, 1970).

In a static inhalation study, mice, rats, rabbits and guinea-pigs were exposed to N-carboxyanthranilic acid anhydride in a 2 m³ chamber. The test substance as a solution in dimethyl sulphoxide and polyethylene glycol 400 (1:1) was nebulized half-hourly during periods of 1 and 4 hours. A propeller ensured uniform distribution in the air. The concentrations measured in the exposure chamber were 49 and 62 mg/m³ air for the 1-hour treatment, and 46 and 130 mg/m³ air for the 4-hour treatment. There were neither signs of intoxication nor mortalities during the 14-day post-treatment observation period (Bayer, 1970).

4-Hour inhalative exposure to 197 mg/m³ air (measured concentration) in a dynamic inhalation chamber on each of 5 successive days produced no symptoms of intoxication in 10 male and 10 female Wistar II rats (Bayer, 1970).

30 Male Wistar II rats were given 200 mg N-carboxyanthranilic acid anhydride/kg body weight orally on the first day of the study. The dose was increased by a factor of 1.5 each time from the 2nd to the 5th day of the study (300 to 1013 mg/kg body weight). There were no mortalities during the 5-day administration time or during the 8-day post-treatment observation period. It was thus not possible to discern any cumulative effect on the part of the substance (Bayer, 1970).

10 Male and 10 female Wistar II rats were exposed in a dynamic inhalation chamber to an average of 11.6 and 71 mg/m^3 air for 4 hours/day on 5 successive days/week for 4 weeks. A control group of 10 male and 10 female animals was exposed to the vehicle (DMSO/Lutrol mixture, 1:1). The exposure concentration in the inhalation chamber was measured spectrophotometrically. The erythrocyte, leucocyte and thrombocyte counts, hemoglobin content, hematocrit, mean corpuscular hemoglobin (Hb_E value), mean corpuscular volume (MCV), prothrombin time, ALAT/GPT activity, ASAT/GOT activity and SDH activity, urea and creatinine content of the serum were examined in 5 male and 5 female rats from each group. Following exposure, the heart, testes, liver, lungs, spleen, adrenals, kidneys, ovaries and thyroids of all the animals were weighed and macroscopically assessed. No histopathological examination was performed. With neither of the concentrations were there any symptoms of intoxication during exposure; body weight gain was within the range observed in the controls. Hematological and biochemical parameters revealed no significant changes. Only the female rats exposed to a concentration of 71 mg N-carboxyanthranilic acid anhydride/m^3 air displayed a significant ($p<0.01$) increase in absolute and relative lung weight and a significant ($p<0.05$) increase in relative liver weight. The remaining organs of both study groups showed no substance-related changes in weight. In both study groups, the hematological and biochemical parameters investigated did not differ from those of the control group. 11.6 mg/m^3 was tolerated without symptoms (Bayer, 1970).

7.3 Skin and mucous membrane effects

Exposing the shaved dorsal skin of white rabbits to a 50% aqueous preparation of N-carboxyanthranilic acid anhydride for 1, 5 and 15 minutes produced no symptoms of irritation (semi-occlusive). Exposure for 20 hours produced a temporary, uncertain reddening. The 50% aqueous preparation proved to be non-irritant to the skin of the ear when this was exposed occlusively for 20 hours. The substance was therefore evaluated as not irritating to the skin (BASF, 1971).

In a further study, the acute skin irritant effect of N-carboxyanthranilic acid anhydride was tested on the ears of rabbits (2 animals). A cotton wool pad containing 50 mg of the test substance was applied to the skin and fixed for a single 24-hour period using an adhesive dressing. No changes could be observed in the treated skin on removing the dressing or during the 7-day post-treatment observation period (Bayer, 1970).

When the undiluted substance was instilled into the eyes of 2 white rabbits a slight reddening and clouding and a severe oedema developed after 1 hour. The symptoms increased after 24 hours with additional bleedings in the conjunctiva, but were declining after 8 days with the exception of a slight reddening of the conjunctiva. Therefore N-carboxyanthranilic acid anhydride was irritant to the eye in this study (BASF, 1971).

In order to test the acute irritant effect on the eye, approx. 50 mg of N-carboxyanthranilic acid anhydride was instilled into the conjunctival sac of 2 rabbits. No changes were observed in the conjunctiva, sclera or cornea (no further information; Bayer, 1970).

7.4 Sensitization

N-Carboxyanthranilic acid anhydride had a sensitizing effect in the maximization test according to Magnussen/Kligman on the skin of guinea-pigs (BASF, 1990 b).

7.5 Subchronic and chronic toxicity

No information available.

7.6 Genotoxicity

7.6.1 In vitro

In the Salmonella/microsome test, there was no evidence of N-carboxyanthranilic acid anhydride, solved in DMSO, having a mutagenic effect in strains TA 1535, TA 100, TA 1537 and TA 98 at doses of 20 to 5000 µg/plate with and without Aroclor 1254 induced S9-mix. A bacteriotoxic effect was observed from 2000 µg/plate (BASF, 1985).

7.6.2 In vivo

5 Male and 5 female NMRI-mice were treated orally by gavage with N-carboxyanthranilic acid anhydride in polyethylene glycol 400, in doses of 200, 670 and 2000 mg/kg (according to EEC directive 79/831, Annex V, B12). After 24 hours, 1000 polychromatic erythrocytes of the bone marrow per animal were examined in all dose groups, after 48 hours in the 2000 mg/kg group only, but in none of the groups or preparation times was the number of micronuclei increased (CCR, 1992).

7.7 Carcinogenicity

No information available.

7.8 Reproductive toxicity

In a fertility study, 10 female Swiss mice were given N-carboxyanthranilic acid anhydride at a concentration of 2% in food 7 days before mating and during the 14-day mating period. With an assumed body weight of 20 g and daily food consumption of 150 g/kg body weight/day, this was equivalent to 3000 mg substance/kg body weight/day. There was a distinct reduction in the number of pregnant mice (20% relative to the controls). A further 10 groups of 9 to 30 female mice were given N-carboxyanthranilic acid anhydride at a concentration of 0.5, 1 and 2% in food 1.5 days before mating as well as during the 3.5-day mating period. This corresponded to 750, 1500 and 3000 mg/kg body weight/day. 10 Groups of 10–20 animals served as controls. In each case 5 females were mated with 1 untreated male. The average fertility index of the control group was 50%, that of the 0.5, 1 and 2% groups 33.3%, 15.4% and 17.4%, respectively. The litter size in the control group was 6.9 animals, that for the substance groups (0.5, 1, 2%) was 6, 5.4 and 4.6 young, respectively. Body weight gain in the dams was within the range observed in the control groups (Cutting et al., 1966).

A further group of 10 female mice were each given 4 mg N-carboxyanthranilic acid anhydride in the morning and 8 mg in the afternoon subcutaneously each day 1.5 days before mating and during the 3.5-day mating period. 15 Animals served as controls. 5 Females were mated with 1 untreated male. The fertility index was 20% (controls 67%), the number of young per litter was 7 (controls 8). Body weight gain in the dams was within the range observed in the control group (Cutting et al., 1966).

7.9 Effects on the immune system

No information available.

7.10 Neurotoxicity

No information available.

7.11 Other effects

10^{-3} M N-Carboxyanthranilic acid anhydride (equivalent to 163 µg/ml) inhibited the concanavalin A and IgE stimulated release of histamine from peritoneal rat mast cells by 90.5% and 98.5%, respectively (Otsuka et al., 1988).

A concentration of 1.25×10^{-4} M N-carboxyanthranilic acid anhydride produced 50% inhibition of the activity of pancreatic elastase, α-lytic protease, α-chymotrypsin, papain and trypsin after <0.25 to 2 minutes (Moorman and Abeles, 1982).

At concentrations of 10^{-4} to 10^{-3} M N-carboxyanthranilic acid anhydride (equivalent to 16.314 to 163.14 µg/ml) it was possible after 30 minutes incubation to observe maximum inhibiting of swine pancreatic elastase acitivity and trypsin activity of 62% and 88%, respectively (Shinogi et al., 1985).

2.5×10^{-6} M N-Carboxyanthranilic acid anhydride (equivalent to 0.4 µg/ml) resulted in 50% inhibition of the bovine serine protease, chymotrypsin, after 2.7 minutes incubation. Bovine trypsin and bovine thrombin were not inactivated by this concentration. The threshold value for 50% inhibition of these enzymes was 25×10^{-6} M (equivalent to 4 µg/ml; Gelb and Abeles, 1986).

8. Experience in humans

The acute skin irritant effect of N-carboxyanthranilic acid anhydride was tested on the forearm of 8 male probands. A cotton wool pad containing approx. 50 mg of the test substance was fixed to the skin for 24 hours using a plaster dressing. No changes could be observed in the treated skin on removing the dressing or during the 7-day post-treatment observation period (Bayer, 1970).

After temporary inhalative exposure of N-carboxyanthranilic acid anhydride (approx. <2 mg/m^3) during production processes of dyes and crop protection chemicals no substance related sensitization effects in the respiratory tract or in the skin were observed in the workers (BASF, 1991).

References

BASF AG, Gewerbehygienisch-Pharmakologisches Institut
Ergebnis der gewerbetoxikologischen Vorprüfung – Isatosäureanhydrid
Unpublished report no. XX 416 (1971)

BASF AG, Abteilung Toxikologie
Report on the study of Isatosäureanhydrid in the Ames test (standard plate test with Salmonella typhimurium)
Unpublished report (1985)

BASF AG
Isatosäureanhydrid
Grunddatensatz für Großstoffe (1987)

BASF AG
Isatosäureanhydrid
Grunddatensatz für Großstoffe (1988)

BASF AG, Abteilung Toxikologie
Study on the acute inhalation toxicity LC_{50} of Isatosäureanhydrid as a dust aerosol in rats, 4-hour exposure
Unpublished report no. 13I0554/897041 (1990 a)

BASF AG, Abteilung Toxikologie
Report on the maximization test for the sensitizing potential of Isatosäureanhydrid in guinea pigs
Unpublished report no. 30H0554/892366 (1990 b)

BASF AG, Abteilung Arbeitsmedizin und Gesundheitsschutz
N-Carboxyanthranilic acid anhydride
Written communication to the BG-Chemie from June 24, 1991

Bayer AG, Institut für Toxikologie
Toxikologische Untersuchungen von Isatosäureanhydrid
Unpublished report no. 2143 (1970)

CCR (Cytotest Cell Research GmbH & Co. KG)
Micronucleus assay in bone marrow cells of the mouse with N-Carboxyanthranilsäureanhydrid
CCR project 264104 (1992)
On behalf of by BG-Chemie, Heidelberg

Cutting, W. C., Rogers, J., Roberts, J., Tabar, P.
Antifertility effects of isatoic anhydride and derivatives
Med. Pharmacol. Exp., 15, 7–16 (1966)

Gelb, M. H., Abeles, R. H.
Substituted isatoic anhydrides: selective inactivators of trypsin-like serine proteases
J.Med. Chem., 29, 585–589 (1986)

Moorman, A. R., Abeles, R. H.
A new class of serine protease inactivators based on isatoic anhydride
J. Am. Chem. Soc., 104, 6785–6786 (1982)

Otsuka, H., Hirai, Y., Nagao, T., Yamasaki, K.
Anti-inflammatory activity of benzoxazinoids from roots of Coix lachrymajobi var. ma-yuen
J. Nat. Prod., 51, 74–79 (1988)

Shinogi, M., Agha, B. J., Tsuji, K., Digenis, G. A.
Assessment of antiprotease activity of some carbamate derivatives
J. Pharm. Sci., 74, 482–485 (1985)

Propionaldehyde

1. Summary and assessment

Propionaldehyde is oxidized by aldehyde dehydrogenases to propionic acid, which can then be channelled into fatty acid metabolism or the citric acid cycle bound to Coenzyme A.

Based on the available animal studies, propionaldehyde is of low to moderate acute toxicity after oral, dermal or inhalation exposure (LD_{50} rat, oral >800 mg/kg body weight; LD_{50} rabbit, dermal >4047 mg/kg body weight; LC_{50} rat, 30 minutes >59800 mg/m^3 air). The observed signs of toxicity are predominantly depression of the central nervous system and, after inhalation exposure, marked irritation of the mucous membranes.

After repeated daily inhalation of 90 ppm (equivalent to 217 mg/m^3) for 6 hours over 4 weeks, rats have shown no signs of toxicity. Continual inhalation of 1 mg/m^3 by rats for 3 months has been tolerated without overt effects.

Neat propionaldehyde can be slightly irritating or corrosive to the skin and eyes of laboratory animals depending on the duration of exposure.

Propionaldehyde appears not to have any mutagenic activity in prokaryotes. In experiments to detect genetic or chromosomal damage in CHO cells, propionaldehyde induces chromosomal aberrations, sister chromatid exchange, and DNA damage (single-strand breaks).

The one study available on the reproductive toxicity of propionaldehyde is of practically no value because of considerable flaws in the protocol.

Propionaldehyde has an indirect sympathomimetic effect on animals.

In man, the application of a 75% aqueous propionaldehyde solution to the skin leads to slight or marked erythema. The olfactory threshold is 0.05 to 1 mg/m^3 air, and the irritant threshold for the eyes and the respiratory tract is 14 to 16 mg/m^3 air.

2. Name of substance

2.1	Usual name	Propionaldehyde
2.2	IUPAC-name	Propanal
2.3	CAS-No.	123-38-6

3. Synonyms, common and trade names

Methylacetaldehyde
Propaldehyde
Propanaldehyde
Propional
Propionaldehyde
Propionic aldehyde
Propylaldehyde
Propylic aldehyde
Propionaldehyd

4. Structural and molecular formulae

4.1 Structural formula

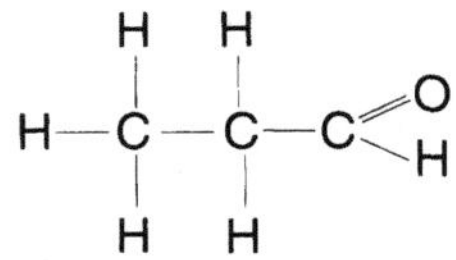

4.2 Molecular formula C_3H_6O

5. Physical and chemical properties

5.1	Molecular mass, g/mol	58.09
5.2	Melting point, °C	–80 (BASF, 1987)
5.3	Boiling point, °C	47–48 (BASF, 1987)
5.4	Vapour pressure, hPa	343 (BASF, 1987)
5.5	Density, g/cm³	0.803 (at 20 °C) (BASF, 1987)
5.6	Solubility in water	280 g/l (at 20 °C) (BASF, 1987) Limited miscibility at room temperature; miscibility increases with decreasing temperature; completely miscible below ca. 15 °C (Falbe and Payer, 1974)

5.7 Solubility in organic solvents	Dissolves completely in ether and alcohol and in many organic solvents (Falbe and Payer, 1974; Weast, 1988)
5.8 Solubility in fat	No information available
5.9 pH-value	–
5.10 Conversion factor	1 ppm ≙ 2.37 mg/m³ 1 mg/m³ ≙ 0.42 ppm (at 25 °C and 1013 hPa)

6. Uses

For the manufacture of propanol and propionic acid, drugs, pesticides, vulcanization accelerators and fragrances (Falbe and Payer, 1974; BASF, 1987).

7. Experimental results

7.1 Toxicokinetics and metabolism

Propionaldehyde has been shown to be oxidized to propionic acid by aldehyde dehydrogenases. During or after oxidation, propionic acid may be bound to Coenzyme A to be channelled into fatty acid metabolism or the citric acid cycle as a substrate, propionyl-CoA (Brabec, 1981).

Rat liver microsomal aldehyde dehydrogenase had a lower affinity for propionaldehyde when compared with related aldehydes. The affinity increased as follows: propionaldehyde < hexanal < heptanal < octanal < nonanal (Antonenkov et al., 1987; Mitchell and Petersen, 1989).

The V_{max} value for propionaldehyde was 6.53 µmol liberated NADH/minute/mg protein, while for nonanal it was 23.0 µmol (Mitchell and Petersen, 1989).

Propionaldehyde reacts with glutathione to form the corresponding thiohemiacetate. Whether this reaction is enzymatic or spontaneous has not yet been determined (no further details; Brabec, 1981).

Table 1. Acute toxicity of propionaldehyde

Species	Number of animals/ dose	Sex	Route	Observation period (days)	LD_{50}/LC_{50} 95% confidence limits	Reference
Mouse	n.s.	n.s.	Oral	n.s.	>800 mg/kg b.w.	Kodak, 1971
Mouse	n.s.	n.s.	i.p.	7	ca. 717 mg/kg b.w.	BASF, 1975
Mouse	8	n.s.	s.c.	21	680 mg/kg b.w.	Skog, 1950
Mouse	n.s.	n.s.	Inhalation (duration of exposure not specified)	n.s.	20 000 mg/m³ air	Wang, 1957
Mouse	n.s.	n.s.	2-hour inhalation	n.s.	21 800 mg/m³ air (17 000–27 900 mg/m³ air)	Izmerov et al., 1982
Rat	n.s.	n.s.	Oral	n.s.	800–1600 mg/kg b.w.	Brabec, 1981
Rat	n.s.	n.s.	Oral	14	1410 mg/kg b.w. (960–2080)	Smyth et al., 1951
Rat	n.s.	n.s.	Oral	7	ca. 3300 mg/kg b.w.	BASF, 1975
Rat	n.s.	n.s.	s.c.	n.s.	800 mg/kg b.w.	Brabec, 1981
Rat	8	n.s.	s.c.	21	820 mg/kg b.w.	Skog, 1950
Rat	n.s.	n.s.	30-minute inhalation	n.s.	26 000 ppm ('62 660 mg/m³ air)	Brabec, 1981
Rat	8	n.s.	30-minute inhalation	21	62 000 mg/m³ air	Skog, 1950

Guinea-pig	n.s.	n.s.	Dermal	n.s.	10 000–20 000 mg/kg b.w.	Brabec, 1981
Rabbit	n.s.	n.s.	Dermal	14	4047 mg/kg b.w. (2730–5990)	Smyth et al., 1951
Rabbit	n.s.	n.s.	Dermal	n.s.	5000 mg/kg b.w.	Brabec, 1981

n.s. = not specified, b.w. = body weight

7.2 Acute and subacute toxicity

The LD_{50} values shown in Table 1 have been reported for oral, intraperitoneal and subcutaneous administration. Signs of toxicity observed included dyspnoea, apathy, atonia, exsiccosis, reeling, lying on the stomach or side, tremors, cessation of breathing and narcosis (BASF, 1975; Owe-Larsson and Skog, 1951; Skog, 1950; Skog, 1952). The surviving animals recovered within 1 to 2 days of administration. Autopsy after oral administration revealed acute heart dilation, particularly on the right side, acute congestive hyperaemia, diffuse redness of the glandular stomach, hardening and thickening of the forestomach wall, diverticula formation and isolated ulceration (BASF, 1975; Skog, 1950). Effects observed after subcutaneous administration included haemorrhages, oedematous vessel walls and slight intraalveolar oedema of the lungs as well as dispersed perivascular oedema in the liver and kidneys (Skog, 1950). There were no treatment-related effects in the mouse after intraperitoneal administration (BASF, 1975).

After inhalation of propionaldehyde, the 2-hour LC_{50} in the mouse was 21 800 (17 000 to 27 900) mg/m^3 air and the 30-minute LC_{50} in the rat was 59 800 or 62 000 mg/m^3 air (see also Table 1). Signs of toxicity observed in mice, rats, guinea-pigs and rabbits included gasping, bradypnoea, clouding of the cornea, severe irritation of the mucous membranes, escape reactions, lying on the back, immobility, stretching spasms and narcosis (BASF, 1975; Gage, 1970; Salem and Cullumbine, 1960; Skog, 1950; Skog, 1952; Wang, 1957).

Post-mortems of the mice, rats, guinea-pigs and rabbits revealed oedema, haemorrhages and hyperaemia in the lungs. In addition dilated alveoli, burst alveolar septa, thickening of the lungs and fluid in the pleural cavity were seen in the mice and the rabbits, while in the rat, bronchitis and bronchopneumonia were observed as well as hyperaemia of the liver and kidneys and acute dilation and congestive hyperaemia of the heart (BASF, 1975; Salem and Cullumbine, 1960; Skog, 1950).

Propionaldehyde decreased the respiratory rate in mice and rats. After 1-minute and 10-minute exposures, the RD_{50} values (reduction of the respiratory rate by 50%) shown in Table 2 were determined.

Table 2. RD_{50} values after exposure to propionaldehyde

Species	RD_{50}	95% Confidence limits
Mouse (Swiss-Webster)	2052 ppm; 10 minutes (4720 mg/m³ air)	1626–3040 ppm (3738–6992 mg/m³ air)
Mouse (Swiss)	2751 ppm; 1 minute (6327 mg/m³ air)	2294–4009 ppm (5276–9921 mg/m³ air)
Mouse (B6C3F1)	2078 ppm; 10 minutes (4780 mg/m³ air)	1803–2402 ppm (4147–5525 mg/m³ air)
Rat (Fischer-344)	6789 ppm; 10 minutes (15 615 mg/m³ air)	3903–15 466 ppm (8977-35 572 mg/m³ air)

A recovery to at least 90% of the initial respiratory rate occurred in all of the studies within 5 minutes of the end of exposure (Alarie, 1981; Babiuk et al., 1985; Steinhagen and Barrow, 1984).

Repeated inhalation of propionaldehyde (dynamic whole-body exposure) at a concentration of 1300 ppm (ca. 3000 mg/m³ air) for 6 hours/day on 6 consecutive days led to a reduction in body weight gain in 4 male and 4 female Alderly Park rats. The behaviour of the rats was normal. There were no treatment-related effects on the haematological and clinical-chemical parameters studied (Hb concentration, haematocrit value, differential blood count, urea, sodium and potassium levels, platelet count and clotting time) or on the urinary parameters (specific gravity, pH, sugar, bilirubin and protein contents). No treatment-related effects were observed macroscopically or microscopically in the lungs, kidneys, spleen, adrenal glands, heart, jejunum, ileum, and thymus. The liver cells were vacuolized (no further details; Gage, 1970).

Continuous inhalation with whole-body exposure at concentrations of 10 and 4 mg/m³ air for 3 weeks caused no changes in behaviour, general well being or body weight gain in male Wistar rats. Post-mortems revealed dystrophic changes in the parenchymatous organs, desquamative bronchitis and, in 30% of all cases, interstitial pneumonia (no further details; Melnikova and Tokanova, 1983).

In a preliminary study, 4 male and 4 female Alderly Park rats were exposed (dynamic whole-body exposure) to propionaldehyde at a concentration of 90 ppm (equivalent to 217 mg/m³ air) for 6 hours/day, 5 days/week for 4 weeks. Body weight gain and behaviour were unaffected. The haematological values (Hb-concentration, haematocrit value, differential blood count, thrombocyte count, urea, sodium and potassium contents, clotting time) and the urinary values (specific gravity, pH, sugar, bilirubin and protein contents) were unchanged. There were also no detectable treatment-related macroscopic or microscopic effects on the lungs, liver, kidneys, spleen, adrenal glands, heart, jejunum, ileum or thymus (Gage, 1970).

Male Wistar rats were continuously exposed to propionaldehyde in a whole-body inhalation chamber at concentrations of 4, 10, 50, 100, 225, 500 or 1300 mg/m^3 for short or long periods. At the higher concentrations (225, 500, 1300 mg/m^3) short-term inhalation (no further details) led to increased motor activity, tachypnoea, irritation of the upper airways and the mucous membranes of the eyes, and subsequently to atonia and decreased motor activity. Long-term exposure (no further details) to 4, 10, 50 or 100 mg/m^3 caused no changes in behaviour or general well-being. Body weight remained unaffected compared with the control group over the entire concentration range (4 to 1300 mg/m^3). Histological examination revealed hyperaemia of the lungs, heart, liver and kidneys and haemorrhages in the lungs in all exposed animals. In addition, concentrations of 4 to 100 mg propionaldehyde/m^3 caused parenchymatous dystrophy in the liver and kidneys. The 1300 mg/m^3 concentration led to dilation and congestion of the veins and capillaries of the interalveolar septa in the lungs, as well as haemorrhagic foci, peribronchial and perivascular oedema, bronchitis and peribronchitis. In the heart, congestion of the veins and capillaries of the interstitial tissues, hyperaemia of the vessels in the subepicardial layer and distended veins were seen at this concentration. These findings were interpreted by the authors as a disturbance of the cardiovascular system (Melnikova and Tokanova, 1983; Tokanova, 1982).

7.3 Skin and mucous membrane effects

The acute skin irritancy of propionaldehyde was studied on the shaved dorsal skin of white rabbits in a patch test, in which a cotton wool patch (ca. 2.5 cm × 2.5 cm) to which the neat substance had been applied was placed in contact with the skin once for 1, 5 or 15 minutes or for 20 hours. After exposure for 1, 5 or 15 minutes, the treated skin was first washed with neat polyethylene glycol 400 and then with a 50% aqueous solution. After the 20-hour exposure, the skin received no further treatment. After exposure for 1 minute, marginal redness was detectable 24 hours later. The 5-minute exposure led to marked reddening, a slight, mottled oedema and marginal necrosis, the 15-minute exposure to mild necrosis. After exposure for 20 hours (occlusive), slight redness and severe oedema were observed. 8 Days later, the skin that had been exposed for 1 minute appeared normal. The skin exposed for 5 or 15 minutes or for 20 hours had mild to marked necrosis and it was parchment-like to leathery in appearance (BASF, 1975).

In a further study, 3 rabbits (white Viennese) had 0.5 ml of neat propionaldehyde applied to the intact, clipped dorsal skin for 3 minutes or 1 hour in a patch test. The treated skin was then washed first with neat polyethylene glycol 400 and then with a 1 : 1 aqueous solution. No effects on the skin were detectable either immediately after exposure or after 30 to 60 minutes, 24, 48 or 72 hours (BASF, 1986 a).

In a study carried out according to OECD guideline no. 404, 0.5 ml neat propionaldehyde was applied semi-occlusively to the intact, clipped dorsal skin of 3 rabbits (white Viennese) for 4 hours in a patch test. After exposure the skin was first washed with neat polyethylene glycol 400, and then with a 1 : 1 aqueous dilution. The observation period was 72 hours. Very slight erythema was detectable in 2/3 rabbits after the patch was removed (4 hours), but this had cleared up completely 24 hours after exposure (BASF, 1986 b).

In another study, the primary skin irritancy in 5 rabbits was classified as grade 2 (slight hyperaemia) out of a maximum of 10 (no further details; Smyth et al., 1951).

The application of 500 mg propionaldehyde to the skin of the rabbit (non-occlusive) caused slight skin irritation (no further details; Union Carbide, 1958).

The application of propionaldehyde to the skin of guinea-pigs caused severe irritation (no further details; Fassett, 1962).

In a study of the acute eye irritancy of propionaldehyde, 50 µl of the neat substance was instilled into the conjunctival sac of the rabbit and carefully distributed over the surface of the eye. After 1 hour, slight redness and clouding of the cornea were observed, as well as secretion and pronounced oedema. 24 Hours later, marked redness, clouding of the cornea and oedema were seen, together with a viscous layer on the surface of the eye and scar formation. 8 Days after instillation, slight redness, marked clouding of the cornea, a partially white nictating membrane and scar formation were still evident (BASF, 1975).

Marked to severe redness and swelling of the conjunctiva were observed in female rabbits (3 white Viennese) 1 hour after instillation of 0.1 ml neat propionaldehyde into the conjunctival sac, in a study conducted according to OECD guideline no. 405. In addition, 24 hours after instillation, slight clouding of the cornea occurred in 2 of the rabbits. A slight retraction of the eye lids and constriction of the pupil occurred 24 hours after instillation in 1 rabbit, and after 48 and 72 hours, respectively, in the other 2. Pus had formed in 2 of the rabbits' eyes after 48 hours. After 8 days, retractions of the eye lids had occurred in all of the rabbits, and in 1 animal slight clouding of the cornea, constriction of the pupil, marginal vascularization of the cornea and damage of the cornea were seen. These effects had still not cleared up in 2/3 rabbits after 21 days, and the clouding of the cornea had become even more severe (grade 2; BASF, 1986 c).

A single instillation of 41 mg propionaldehyde into the conjunctival sac of the rabbit caused severe eye irritation (no further details; Union Carbide, 1958).

In another study in rabbits, examination of the eyes 24 hours after instillation of 20 µl neat propionaldehyde into the conjunctival sac revealed severe irritation and necrosis (irritation grade 5 out of a maximum of 10; no further details; Smyth et al., 1951).

7.4 Sensitization

No information available.

7.5 Subchronic and chronic toxicity

Male Wistar rats were continuously (no further details) exposed to propionaldehyde for 3 months at concentrations of 0.01, 0.1 or 1.0 mg/m^3 in a whole-body inhalation chamber. The body weights and behaviour of the rats were unchanged relative to the controls. Histologically, no morphological changes were evident in the internal organs compared with the control group (no further details; Melnikova and Tokanova, 1983).

7.6 Genotoxicity

7.6.1 In vitro

Propionaldehyde (97% pure) was not mutagenic in the *Salmonella typhimurium* strains TA 98, TA 100, TA 1535 and TA 1537 in the direct plate test in two independent assays (3 plates per concentration) at concentrations of 100 to 10 000 μg/plate, with and without metabolic activation (S9-mix from Aroclor 1254-induced rat or hamster liver). The high concentrations were cytotoxic (Mortelmans et al., 1986).

There were also no indications of mutagenic activity in 5 other studies in the *Salmonella typhimurium* strains TA 98, TA 100, TA 102, TA 1535 and TA 1537 with or without metabolic activation. No information was given on whether the cytotoxic range was reached (Aeschbacher et al., 1989; Florin et al., 1980; NTP, 1983; Pool and Wiessler, 1981; Sasaki and Endo, 1978).

In another study to determine the cytogenetic effect, propionaldehyde led to an increase in the chromosomal aberration rate in CHO cells (Chinese hamster ovary cells), and was evaluated as a mutagen (no further details; NTP, 1985).

The sister chromatid exchange rate in human lymphocytes after exposure to propionaldehyde was within the control range. Concentrations of 2×10^{-3} and 3×10^{-3} % (v/v, equivalent to 16.1 and 24.1 μg/ml, respectively) were used, with exposure times of 24 and 48 hours. The number of metaphases evaluated per concentration was 32 to 100, and the assay was only carried out without metabolic activation. It was not specified whether the cytotoxic range was reached (Obe and Beek, 1979).

The sister chromatid exchange rate was increased in Chinese hamster ovary cells, and the substance was therefore considered to be genotoxic (no further details; NTP, 1985).

In an alkaline elution test, propionaldehyde caused a concentration-dependent increase in DNA single-strand breaks and a reduction in cross-links in CHO-K_1 cells, from a concentration of 0.5 mM (equivalent to 29.1 μg/ml; Marinari et al., 1984).

7.6.2 In vivo

No information available.

7.7 Carcinogenicity

No information available.

7.8 Reproductive toxicity

The embryos of Sprague-Dawley rats were given intra-amniotic injections of 10, 100 or 1000 μg propionaldehyde in utero on day 13 of pregnancy. The contralateral embryos remained untreated. A control group was treated with physiological saline. 5 To 6 dams were used per treatment group, and 18 were used in the control group. The numbers of malformations and of dead and resorbed foetuses were recorded on day 20 of pregnancy. Propionaldehyde caused a dose-dependent increase in the mortality and resorption rates, which was ca. 50% at 1000 µg/foetus ($p<0.05$). A dose-dependent increase in dead and resorbed foetuses (ca. 24% in the 1000 µg group) also occurred in the corresponding contralateral uterine sides, but the value was still within the control range. In the control group the mortality and resorption rate was ca. 24% (injection of physiological saline), and in the corresponding contralateral uterine side it was 12%. Only 1 of the foetuses treated with 1000 µg was malformed (shortened tail; number of foetuses examined not specified). In the control group, 6% of the foetuses treated with physiological saline and 3% of the untreated contralateral foetuses were malformed (type of malformations not specified; Slott and Hales, 1985). When considered in terms of current requirements for reproductive studies, this study is of limited value.

7.9 Effects on the immune system

No information available.

7.10 Neurotoxicity

No information available.

7.11 Other effects

Doses of 5 or 10 mg propionaldehyde/kg body weight given to rats by intravenous injection caused a dose-dependent increase in blood pressure of up to 12.4%, while 20 and 40 mg/kg body weight caused a fall of up to 63.9% (Egle et al., 1973; Green and Egle, 1983).

After inhalation of 10 000 to 200 000 mg propionaldehyde/m^3 air (equivalent to 4200 to 84 000 ppm) for 1 minute there was a significant dose-dependent increase in blood pressure of up to 47% in rats (Egle, 1972).

Rabbits and cats also showed a dose-dependent increase in blood pressure after intravenous injections of 5 to 50 mg propionaldehyde/kg body weight, and a fall by ca. 50% at 58 to 80 mg/kg body weight (no quantification; Skog, 1952).

The blood pressure increased in dogs given an intravenous injection of 3.6 to 31.3 mg propionaldehyde/kg body weight (no quantification; Romano et al., 1954; Wingard et al., 1955).

Rats showed a 71.2% decrease in heart rate after intravenous administration of 40 mg propionaldehyde/kg body weight (Egle et al., 1973).

In dogs, the heart rate was unaffected after intravenous injection of 5.8 mg propionaldehyde/kg body weight (Romano et al., 1954).

The heart rate of rats was not significantly affected after inhalation of 3000 to 10 000 mg propionaldehyde/m^3 air for 1 minute. Concentrations of 20 000 to 30 000 mg/m^3 led to a significant increase ($p<0.01$) by up to 6.1%, 50 000 to 150 000 mg/m^3 air had no significant effect, and 200 000 mg/m^3 air led to a significant decrease ($p<0.01$) in the heart rate of 26.0% (Egle, 1972).

Intravenous injection of phentolamine, propranolol or atropine prior to propionaldehyde administration in rats was antagonistic to both the increase and decrease in blood pressure and to the decrease in heart rate (Egle et al., 1973).

The threshold concentration for a local anaesthetic effect in guinea-pigs after intradermal injection of propionaldehyde was 0.01 M (equivalent to 600 mg/l). Marked irritation was seen at the injection site (no further details; Romano et al., 1954).

In isolated Langendorff rabbit hearts, 5×10^{-5} M propionaldehyde (equivalent to 2.9 μg/ml) was positively chronotropic and inotropic, and the heart rate increased by 22% (no details of inotropism; Romano et al., 1954).

Propionaldehyde had a mild spasmolytic effect in isolated rabbit duodenum at 5×10^{-5} M (equivalent to 2.9 μg/ml), and at 5.48×10^{-6} M (equivalent to 0.32 μg/ml) was mildly tocolytic (i.e. inhibited contractions) in rat uterus (Romano et al., 1954).

A 1% aqueous propionaldehyde solution caused 75% haemolysis in rabbit blood in vitro, while in vivo, subcutaneous injection with 100 to 500 mg propionaldehyde/kg was not haemolytic in 8 rats (no further details; Skog, 1952).

The average toxicity of propionaldehyde to 4 cell cultures was classified as 2.3 on a scale from 0 to 9. In detail, the following results were obtained: inhibition of cell growth in a stationary suspension culture of Ascites sarcoma BP-8 cells by a 1 mM propionaldehyde solution (equivalent to 58.09 μg/ml) after 48 hours: 90 to 99%; inhibition of oxidative metabolism in hamster brown adipose cells by a 1 mM solution (equivalent to 58.09 μg/ml) after 5 minutes: 0 to 9%; membrane damage in human lung fibroblasts by a 25 mM solution (equivalent to 1452.25 μg/ml) after 30 minutes: 0 to 9%; inhibition of ciliary activity in chicken embryo tracheal cells by a 5 mM solution (equivalent to 290.45 μg/ml) for 60 minutes: 0 to 9% (Curvall et al., 1984). Inhibition of ciliary activity occurred only after more than 60 minutes (no further details; Pettersson et al., 1982).

8. Experience in humans

The olfactory threshold for propionaldehyde has been given as 1 mg/m^3 (Izmerov et al., 1982) and 0.048 mg/m^3 (Tokanova, 1982).

Irritation of the eyes and upper airways began at 14 to 16 mg/m^3 in human volunteers (Izmerov et al., 1982).

Propionaldehyde was patch-tested in 12 volunteers of Oriental extraction. A filter paper was treated with 25 μl of a 75% aqueous propionaldehyde solution and then applied to the volar forearm, which had previously been soaked in warm water at 33 °C for 10 minutes. The patch was covered and then removed after 5 minutes. The skin reaction was observed for 60 minutes. All 12 volunteers had erythema, the severity of which was assessed by laser-doppler velocimetry: grade 1 – high flushing intensity, grade 25 – low flushing intensity. 5 Volunteers suffered severe erythema (grade 1 to 12), the remaining 7 mild erythema (grade 14 to 25; Wilkin and Fortner, 1985).

Propionaldehyde is a constituent of cigarette smoke, which contains 30 to 940 ng propionaldehyde per ml (Auerbach et al., 1977). It is able to react with human haemoglobin to form an imidazolidinone derivative which can be separated chromatographically; after incubation with cigarette smoke, however, a corresponding propionaldehyde derivative could not be detected unequivocally (Hoberman and San George, 1988).

9. Threshold limit values

No validated threshold limit values have been identified.

Based on a limited 4-week inhalation study in rats, a provisional threshold level of 200 ppm (equivalent to 474 mg/m^3 air) has been suggested for the workplace (Gage, 1970).

References

Aeschbacher, H.U., Wolleb, U., Löliger, J., Spadone, J.C., Liardon, R.
Contribution of coffee aroma constituents to the mutagenicity of coffee
Food Chem. Toxicol., 27, 227–232 (1989)

Alarie, Y.
Dose-response analysis in animal studies: prediction of human responses
Environ. Health Perspect., 42, 9–13 (1981)

Antonenkov, V.D., Pirozhkov, S.V., Panchenko, L.F.
On the role of microsomal aldehyde dehydrogenase in metabolism of aldehydic products of lipid peroxidation
FEBS Lett., 224, 357–360 (1987)

Auerbach, C., Moutschen-Dahmen, M., Moutschen, J.
Genetic and cytogenetical effects of formaldehyde and related compounds
Mutat. Res., 39, 317–362 (1977)

Babiuk, C., Steinhagen, W.H., Barrow, C.S.
Sensory irritation response to inhaled aldehydes after formaldehyde pretreatment
Toxicol. Appl. Pharmacol., 79, 143–149 (1985)

BASF AG, Gewerbehygiene und Toxikologie
Propionaldehyd, Ergebnis der gewerbetoxikologischen Vorprüfung
Unpublished report XXIV 523 (1975)

BASF AG, Abteilung Toxikologie
Report on the acute dermal irritation/corrosivity to the intact dorsal skin of the white rabbit (short-term test)
Unpublished report (1986 a)

BASF AG, Abteilung Toxikologie
Report on the acute dermal irritation/corrosivity to the intact dorsal skin of the white rabbit based on OECD
Unpublished report (1986 b)

BASF AG, Abteilung Toxikologie
Report on the acute irritation to the eye of the white rabbit based on OECD
Unpublished report (1986 c)

BASF AG
Propionaldehyd, Grunddatensatz für Großstoffe, im Anhang DIN-Sicherheitsdatenblatt (1987)

Brabec, M.J.
Aldehydes and acetals (chapter 37)
In: Patty's Industrial Hygiene and Toxicology
Clayton, G.D., Clayton, F.E. (eds.), 3rd ed., vol. 2A, 2629–2669
John Wiley & Sons, New York (1981)

Curvall, M., Enzell, C.R., Pettersson, B.
An evaluation of the utility of four in vitro short term tests for predicting the cytotoxicity of individual compounds derived from tobacco smoke
Cell Biol. Toxicol., 1, 173–193 (1984)

Egle, J.L., jr.
Effects of inhaled acetaldehyde and propionaldehyde on blood pressure and heart rate
Toxicol. Appl. Pharmacol., 23, 131–135 (1972)

Egle, J.L., jr., Hudgins, P.M., Lai, F.M.
Cardiovascular effects of intravenous acetaldehyde and propionaldehyde in the anesthetized rat
Toxicol. Appl. Pharmacol., 24, 636–644 (1973)

Falbe, J., Payer, W.
Aldehyde, aliphatische
In: Ullmann's Encyklopädie der technischen Chemie, 4th ed., vol. 7, 119, 128–129 (1974)
Verlag Chemie, Weinheim

Fassett, D.W.
Aldehydes and Acetals
In: Patty's Industrial Hygiene and Toxicology
Fassett, D.W., Irish, D.D. (eds.), vol. II, 2nd ed., p. 1959–1989 (1962)
New York/London, John Wiley & Sons

Florin, I., Ruthberg, L., Curvall, M., Enzell, C.R.
Screening of tobacco smoke constituents for mutagenicity using the Ames' test
Toxicology, 18, 219–232 (1980)

Gage, J.C.
The subacute inhalation toxicity of 109 industrial chemicals
Br. J. Ind. Med., 27, 1–18 (1970)

Green, M.A., Egle, J.L., jr.
Effects of intravenous acetaldehyde, acrolein, formaldehyde and propionaldehyde on arterial blood pressure following acute guanethidine treatment
Res. Commun. Chem. Pathol. Pharmacol., 40, 337–340 (1983)

Hoberman, H.D., San George, R.C.
Reaction of tobacco smoke aldehydes with human haemoglobin
J. Biochem. Toxicol., 3, 105–119 (1988)

Izmerov, N.F., Sanotsky, I.V., Sidorov, K.K.
Toxicometric parameters of industrial toxic chemicals under single exposure
Centre of International Projects, GKNT, Moscow, p. 102 (1982)

Kodak
Kodak Company Reports (1971)
Cited in NIOSH (1989)

Marinari, U.M., Ferro, M., Sciaba, L., Finollo, R., Bassi, A.M., Brambilla, G.
DNA-damaging activity of biotic and xenobiotic aldehydes in Chinese hamster ovary cells
Cell Biochem. Funct., 2, 243–248 (1984)

Melnikova, A.P., Tokanova, S.E.
Die biologische Wirkung von Propionaldehyd und Propionsäure auf den Organismus von Versuchstieren (translation from the Russian)
Gig. Sanit., Issue 4, 74–76 (1983)

Mitchell, D.Y., Petersen, D.R.
Oxidation of aldehydic products of lipid peroxidation by rat liver microsomal aldehyde dehydrogenase
Arch. Biochem. Biophys., 269, 11–17 (1989)

Mortelmans, K., Haworth, S., Lawlor, T., Speck, W., Tainer, B., Zeiger, E.
Salmonella mutagenicity tests: II. Results from the testing of 270 chemicals
Environ. Mutagen., 8, 1–119 (1986)

NIOSH (National Institute of Occupational Safety and Health)
Registry of Toxic Effects of Chemical Substances (1989)

NTP (National Toxicology Program)
Annual Plan for Fiscal Year 1983
Test results for mutagenicity in Salmonella assays in FY 1982
U.S. Department of Health and Human Services, p. 65 (1983)

NTP (National Toxicology Program)
Annual Plan for Fiscal Year 1985
Chemical test results for cytogenic effects in Chinese hamster ovary cells in FY 1984
U.S. Department of Health and Human Services, p. 86 (1985)

Obe, G., Beek, B.
Mutagenic activity of aldehydes
Drug Alcohol. Depend., 4, 91–94 (1979)

Owe-Larsson, A., Skog, E.
The effect of tetraethylthiuramdisulphide (disulfiram) upon the toxicity of lower aliphatic aldehydes (formaldehyde, acetaldehyde, propionaldehyde, butyraldehyde, acrolein and crotonaldehyde)
Acta Pharmacol. Toxicol., 7, 220–226 (1951)

Pettersson, B., Curvall, M., Enzell, C.R.
Effects of tobacco smoke compounds on the ciliary activity of the embryo chicken trachea in vitro
Toxicology, 23, 41–55 (1982)

Pool, B.L., Wiessler, M.
Investigations on the mutagenicity of primary and secondary α-acetoxynitrosamines with Salmonella typhimurium: activation and deactivation of structurally related compounds by S-9
Carcinogenesis, 2, 991–997 (1981)

Romano, C., Meyers, F.H., Anderson, H.H.
Pharmacological relationship between aldehydes and arterenol
Arch. Int. Pharmacodyn., 99, 378–390 (1954)

Salem, H., Cullumbine, H.
Inhalation toxicities of some aldehydes
Toxicol. Appl. Pharmacol., 2, 183–187 (1960)

Sasaki, Y., Endo, R.
Mutagenicity of aldehydes in Salmonella
Mutat. Res., 54, 251–252 (1978)

Skog, E.
A toxicological investigation of lower aliphatic aldehydes. I. Toxicity of formaldehyde, acetaldehyde, propionaldehyde and butyraldehyde; as well as of acrolein and crotonaldehyde
Acta Pharmacol. Toxicol., 6, 299–318 (1950)

Skog, E.
Anaesthetic and haemolytic action of lower aliphatic aldehydes and their effect on respiration and blood pressure
Acta Pharmacol. Toxicol., 8, 275–289 (1952)

Slott, V.L., Hales, B.F.
Teratogenicity and embryolethality of acrolein and structurally related compounds in rats
Teratology, 32, 65–72 (1985)

Smyth, H.F., jr., Carpenter, C.P., Weil, C.S.
Range finding toxicity data: list IV
Arch. Ind. Hyg. Occup. Med., 4, 119–122 (1951)

Steinhagen, W.H., Barrow, C.S.
Sensory irritation structure-activity study of inhaled aldehydes in B6C3F1 and Swiss-Webster mice
Toxicol. Appl. Pharmacol., 72, 495–503 (1984)

Tokanova, S.E.
Biologische Wirkung und gesundheitliche Beurteilung von Propionaldehyd und Propionsäure als Verschmutzer der Luft in Wohngebieten (translation from the Russian)
Gig. Sanit., Issue 4, 10–13 (1982)

Union Carbide Co.
Union Carbide Data Sheet (1958)
Cited in NIOSH (1989)

Wang, W.Y.
Toxicology of fatty aldehydes
Sbornick Rabot Toksikol. Lab., Gosudarst, Nauch.-Issledovatel. Inst. Gig. Tr. Prof. Zabol., 6, 42–60 (1957)
From: Chem. Abstr., 55, 8653 e (1961)

Weast, R.C. (ed.)
CRC-Handbook of Chemistry and Physics, 67th ed.
CRC-Press, Boca Raton, Florida (1988)

Wilkin, J.K., Fortner, G.
Cutaneous vascular sensitivity of lower aliphatic alcohols and aldehydes in Orientals
Alcohol. Clin. Exp. Res., 9, 522–525 (1985)

Wingard, C., Hitchcock, P., Teague, R.S.
A survey of aldehydes with respect to their action on the blood pressure
Arch. Int. Pharmacodyn. Ther., 102, 65–84 (1955)

Azodicarbonamide

1. Summary and assessment

After intratracheal instillation or inhalation exposure, up to about 90% of azodicarbonamide is absorbed by rats, while after oral administration (stomach tube) only 30% is absorbed. Within minutes, azodicarbonamide is converted to diurea, its only metabolite, and within 72 hours it is excreted, essentially in the urine.

After a single oral or dermal application, azodicarbonamide is practically non-toxic, and after a single inhalation exposure, slightly toxic (LD_{50} rat oral >2500 mg/kg, LD_{50} mouse oral >10 000 mg/kg, LD_{50} rat dermal >500 mg/kg, LC_{50} rat inhalation >6100 mg/m^3).

Repeated oral administration to rats causes slightly decreased iodine uptake by the thyroid gland, but no associated tissue changes. Repeated intraperitoneal administration causes increased mortality and haematuria, but thyroid function is unaffected.

The substance is not a skin irritant in rabbits, but causes reversible swelling and redness of the conjunctiva in the rabbit eye due to mechanical irritation. Effects on the cornea are not observed.

Azodicarbonamide is not a skin sensitizer in guinea-pigs. A 4-week inhalation exposure has caused neither specific nor non-specific respiratory sensitization.

Deaths have been caused by the subchronic oral administration of azodicarbonamide to rats for 90 days, at dose levels of 2500 and 5000 mg/kg body weight/day. Histopathologically, kidney damage has been observed. The no effect level has been between 500 and 100 mg/kg body weight/day in the males, and 1000 mg/kg body weight/day in the females. In contrast, no treatment-related effects have been observed after subchronic oral administration of 1250 mg/kg body weight/day to male mice and 2500 mg/kg body weight/day to female mice. No histopathological effects have been observed in mice and rats after a 13-week inhalation exposure. The no effect level has been 200 mg/m^3.

The compound gives positive results in the Ames test and negative results in *Escherichia coli*. In vivo, it is without effect in the Drosophila test and the mouse micronucleus test.

Dietary administration to rats and dogs of flour that had been treated with azodicarbonamide (ca. 100 mg/kg) during the baking process produced no clinical symptoms or treatment-related organ changes after 2 years of treatment. The tumour incidence was not increased.

In a 3-generation study in male and female rats in which the same treated flour was administered, azodicarbonamide had no effect on fertility, reproduction or lactation in any generation.

The results of these studies with the special diet are only of limited value, because the effects are not necessarily directly attributable to azodicarbonamide.

No indication of effects on sperm morphology or vaginal cytology were observed in a 90-day inhalation study in rats and mice.

In humans, contact with azodicarbonamide causes sensitization effects on the respiratory tract and the skin, and slight changes in the respiratory volume.

2. Name of substance

2.1	Usual name	Azodicarbonamide
2.2	IUPAC-name	Diazenedicarboxamide
2.3	CAS-No.	123-77-3

3. Synonyms, common and trade names

ADA
ADC
Azobiscarbonamide
Azobiscarboxamide
Azodicarbondiamide
Azodicarboxamide
Azodiformamide
1,1'-Azobis formamide
Aceto AD
Cellmic C
Celogen AZ
Diazenedicarbonic acid amide
Genitron AC
Porofor ADC
Vinyfor AC
Diazendicarbonsäureamid

4. Structural and molecular formulae

4.1 Structural formula $H_2N-C(=O)-N=N-C(=O)-NH_2$

4.2 Molecular formula $C_2H_4N_4O_2$

5. Physical and chemical properties

5.1	Molecular mass, g/mol	116.98
5.2	Melting point, °C	>210 (decomposition) (Bayer, 1989)
5.3	Boiling point, °C	No information available
5.4	Vapour pressure, hPa	2.53×10^{-10} (at 20 °C) 3.16×10^{-8} (at 50 °C) (Bayer, 1989)
5.5	Density, g/cm^3	1.65 (at 20 °C) (Bayer, 1989)
5.6	Solubility in water	<0.1 g/l (at 20 °C) (Bayer, 1989)
5.7	Solubility in organic solvents	Soluble in N,N-dimethylformamide (Bayer, 1989)
5.8	Solubility in fat	No information available
5.9	pH-value	Neutral (Bayer, 1989)
5.10	Conversion factor	1 ppm $\triangleq$ 4.77 mg/m^3 1 mg/m^3 $\triangleq$ 0.21 ppm (at 25 °C and 1013 hPa)

6. Uses

A blowing agent for vulcanizates and a leavening agent for baked goods (Eichenhofer and Roos, 1977).

7. Experimental results

7.1 Toxicokinetics and metabolism

Groups of 3 male F-344/N rats were given azodicarbonamide (ADC) as an aqueous suspension by stomach tube at doses of 11.7, 117 and 1170 mg/kg body weight. In addition, a further group was given the highest dose of ADC as a suspension in corn oil.

Less than 10% of the chemical was excreted in the urine and, whether suspended in water or corn oil, nearly all of the remaining dose was eliminated in the faeces within 24 hours (International Research and Development Corporation, 1982 a).

Toxicokinetics and metabolism were studied in male F-344/N Fischer rats (age 11 to 15 weeks, weight 200 to 250 g) after oral (gavage), intratracheal or inhalation exposure (duration of exposure 3, 5.5 or 6 hours). 3 Rats were given 0.1 mg ^{14}C-ADC (5 μCi) as a single dose by stomach tube. 3 Further rats were given the same dose intratracheally. For the 3-hour inhalation exposure, 5 rats were exposed to 1.5 mg ^{14}C-labelled ADC/m^3 air and 3 rats were exposed to 150 mg ^{14}C-labelled ADC/m^3 air. For the 5.5- and 6-hour exposures, the concentration used was 20 mg ^{14}C-labelled ADC/m^3 air. Exposure was carried out in a nose-only inhalation chamber. Directly after exposure, the rats were placed in metabolism cages and their urine, faeces and exhaled air were collected. The ^{14}C-radioactivity in the excretions, tissue samples and exhaled air was measured in a liquid scintillator after the appropriate processing. The metabolites were separated and determined by means of HPLC. In the 5.5-hour inhalation exposure study, 3 rats were taken out of the study 20, 40, 60, 90, 120, 180, 240, 300 and 330 minutes after the beginning of exposure and ^{14}C-labelled ADC was measured in the blood and lungs. These rats (27 in total) were employed as a satellite group. In the 6-hour inhalation exposure, satellite groups each consisting of 3 rats (36 rats in total) were dissected from 15 minutes to 102 days after the end of exposure and the content and concentration of radioactivity in the tissues were measured. The results showed that after gavage or intratracheal administration of ^{14}C-labelled ADC, a metabolite, bisurea (1,2-hydrazine dicarboxamide) was found. In parallel studies in which blood samples from control rats were used it was shown that the transformation of ADC to bisurea occurred within 5 minutes. Saturation of metabolism could only be achieved at very high doses (5 mg ADC/ml). In addition to the metabolite, the parent compound was also found. After administration by stomach tube, the radioactivity was excreted within 72 hours, essentially in the faeces (67%) and the urine (30%). A smaller proportion (ca. 2%) was exhaled as radiolabelled CO_2. After intratracheal administration, the radioactivity was eliminated mainly in the urine (88%) and to a lesser extent in the faeces (ca. 10%). The proportion that was eliminated in exhaled air was 0.8%. A negligible amount of radioactivity was still present in the tissues at this time. In all 4 inhalation exposure studies, the radioactivity was excreted within 72 hours, predominantly in the faeces (58 to 84%) and to a lesser extent in the urine (11 to 34%). Analysis of the blood samples that were taken at regular intervals during the 5.5-hour inhalation exposure revealed the rate of build-up of ^{14}C-labelled ADC was a linear function. This was also established in the lungs of the same animals. Consequently, the ratio of ^{14}C-labelled ADC in the blood to that in the lungs was a linear function with a slope indistinguishable from zero. The time course of absorption and excretion of ^{14}C-labelled ADC in the 6-hour inhalation exposure study (with subsequent monitoring up to day 102 after exposure) resulted in a similar distribution-excretion rate for ADC, i.e. the radioactivity was essentially eliminated within 72 hours (Mewhinney et al., 1987).

F344/N rats (age 7 weeks, sex unspecified) inhaled azodicarbonamide aerosols daily for 6 hours/day, 5 days/week for 13 weeks at the following concentrations: 0 (control), 50, 100 and 200 mg/m^3. The average aerosol concentrations during the 13-week study were 49.8±5.2, 100±7 and 204±10 mg azodicarbonamide/m^3 (mean value ± standard deviation). At the end of the study, the lungs (including parenchyma and bronchi), upper respiratory tract (larynx, trachea and nasal cavities), kidneys, bladder (without contents) and stomach (without contents) were removed and analysed. The aim of this study was to measure azodicarbonamide and its metabolite, diurea, in the tissues. A non-linear dose-dependent amount of diurea was detected in the lungs, while azodicarbonamide was undetectable. However, even at the highest exposure concentration, the amount of diurea in the lungs was less than 1% of the estimated amount of azodicarbonamide retained over the study period. In summary, the authors established that azodicarbonamide is rapidly eliminated from the lungs, even at inhaled concentrations of 200 mg/m^3. Neither azodicarbonamide nor diurea were found in the kidneys at any of the test concentrations after 13 weeks. From these results the authors concluded that the clearance of azodicarbonamide is slightly delayed with increasing dose (Bechtold et al., 1989; Medinsky et al., 1990).

7.2 Acute and subacute toxicity

In material safety data sheets, oral LD_{50} values for rats of >6800 and >6400 mg/kg have been given (Uniroyal; National Polychemical; both documents undated). For the rat and the mouse, the oral LD_{50} value is more than 10 000 mg/kg (International Research and Development Corporation, 1991).

The acute oral toxicity of a mixture of ADC (90%) and zinc salt (10%) has been described in detail. Groups of 15 male Wistar-II rats (weight 160 to 210 g) were given this mixture by stomach tube as an aqueous emulsion in Cremophor EL. The following doses were used: 500, 1000 and 2500 mg mixture/kg body weight. The observation period was 14 days. No effects were observed in the rats at the 500 mg/kg dose. In the rats treated with the higher doses, there was a reduction in well-being for 2 to 6 days, then the rats appeared normal again. Thus in this study an LD_{50} value for the mixture of >2500 mg/kg body weight was established (Bayer, 1972).

The acute oral toxicity of another product, Porofor ADC (ADC content not specified) was determined in rats. Groups of 10 rats (sex, age and weight not specified) were given a single dose of Porofor ADC in polyethylene glycol 400 by stomach tube. The observation period was 7 days. The following doses were administered: 0.5, 1 and 2.5 g/kg body weight. Signs of toxicity included spasms and general malaise. These symptoms occurred at the two higher dose levels, 1 to 2 hours after administration, but were reversible within 24 hours. The median lethal dose for Porofor ADC after oral administration to rats was over 2.5 g/kg body weight (Bayer, 1965).

No signs of toxicity were observed in 10 male Wistar-II rats (160 to 180 g) after a single oral dose of 5.0 g Porofor ADC/K/kg body weight (in Lutrol) during an observation period of 14 days. The LD_{50} was more than 5.0 g/kg body weight (Bayer, 1976).

A single dose of 4.0 g azodicarbonamide/kg body weight (25% aqueous suspension in Dispersol O.G.) was administered to 3 male and 3 female young adult rats (Alderley Park) by stomach tube. The observation period was 7 days. No clinical signs of toxicity or histopathological changes were observed. The oral LD_{50} was thus more than 4.0 g/kg body weight (ICI, 1966).

2 Rabbits were each given 500 mg azodicarbonamide/kg in Lutrol by stomach tube. The following liver function tests were carried out after 1 and 24 hours, and after 7 days: bromosulphthalein test by the Hofmann-Oettel method, determination of serum glutamate-pyruvate-transaminase (SGPT) and serum sorbitol dehydrogenase (SDH). There were no adverse effects on liver function in this study (Bayer, 1965).

Azodicarbonamide was applied to the shaved backs of 5 male rats as a 25% solution in Lutrol (500 mg/kg body weight). The rats were prevented from licking the substance off by the use of paper collars. The substance was not removed from the skin during the 7-day observation period. None of the rats showed clinical signs of toxicity. The substance was thus not absorbed through the intact skin of rats in detectable amounts. The dermal LD_{50} for rats was therefore >500 mg/kg (Bayer, 1965).

The only information available on the acute dermal toxicity of the chemical relates to a mixture (90% ADC, 10% zinc salt). The mixture was applied to the dorsal skin (which had been shaved on the previous day) of 5 male Wistar rats as a 25% aqueous emulsion, at a dose of 500 mg/kg body weight. It was not removed from the skin for 7 days. No overt signs of toxicity were observed in the rats during the study. Thus the dermal LD_{50} value was more than 500 mg/kg body weight (Bayer, 1972).

Central nervous system depression, diarrhoea, cyanosis and dyspnoea occurred in mice given single intraperitoneal injections of 250 mg/kg body weight and above. Deaths were observed from 750 mg/kg body weight (no further details; Oser, 1965).

Groups of 3 male and 3 female Alderley Park rats were given single intraperitoneal injections of azodicarbonamide (as a 25% aqueous suspension in Dispersol O.G.) at doses of 125, 250, 500, 1000 and 2000 mg/kg body weight. The observation period was 7 days. All of the animals in the highest dose group showed general debility within 2 hours of the injection and all died within 24 hours. Macroscopic examination revealed general congestion. Rats that received 1000 and 500 mg/kg likewise showed debility and ruffled fur. After 24 hours the rats that had received 1000 mg/kg were in such poor condition that they had to be killed. Macroscopic examination of these animals revealed congestion and hydrothorax. The kidneys appeared pale and enlarged, and had a soft consistency. 3 Of the 6 rats showed elevated blood urea. Histologically, the kidneys showed pronounced distension of the proximal and distal tubules and crystal deposition. In addition, lympholysis of the thymus was observed and, in the males, inflammatory effects in the epididymides and vas deferens were evident. At a dose of 500 mg/kg, the rats produced a reduced volume of urine, and they died after 2 days. Macroscopically, swollen and oedematous kidneys with crystalline deposits and inflammatory effects were found, as well as atrophied thymus glands. The rats in the two lower dose groups also showed signs of debility. The kidneys were pale with white marks and showed pyelonephritic effects. In order to monitor the level of urea in the blood for 3 days, further groups of 2 male rats were given intraperitoneal injections of 40, 25 or 12.5 mg azodicarbonamide/kg body weight. The blood urea level was within the normal range. Histopathologically, no effects on the kidneys were seen at these doses. In this study, the intraperitoneal LD_{50} value for male rats was between 125 and 250 mg/kg and for females was between 250 and 500 mg/kg (ICI, 1966).

Azodicarbonamide in Lutrol was given to groups of 10 male rats of doses at 10 to 1500 mg/kg body weight as a single intraperitoneal injection. Only at the lowest dose were no signs of toxicity seen, while at all the other dose levels signs of toxicity such as accelerated breathing, reeling, spasticity, somnolence and general malaise were seen after 1 to 2 hours. After 2 to 3 days the symptoms had cleared up. Deaths occurred within 1 to 4 days at doses of 250 mg/kg and above. All of the rats at the highest dose level died. An LD_{50} of 440 mg/kg was calculated from these results (Bayer, 1965).

The intraperitoneal LD_{50} in mice is between 750 and 1250 mg/kg body weight. Symptoms observed included diarrhoea, cyanosis, dyspnoea and depression (Oser et al., 1965).

To study the acute inhalation toxicity, 5 male and 5 female rats underwent whole-body exposure to an average concentration of 6100 mg ADC/m^3 air for 4 hours in a glass chamber. The particle size was 5.8 µm with a standard deviation of 2.25 (determined by cascade impactor). The observation period was 14 days. In 8/10 rats, dyspnoea was observed immediately after the end of exposure. There were no other clinical or histopathological effects. Thus the 4-hour LC_{50} in rats was more than 6100 mg/m^3 air (International Research and Development Corporation, 1982 b).

To study the inhalation toxicity of a mixture of ADC (90%) and zinc salt (10%), 1 g of the mixture was heated in a 0.4 m^3 chamber in which the experimental animals were placed (1st study: 10 male rats and 20 male mice; 2nd study: 5 male hamsters and 2 male rabbits). In this way, an explosive reaction occurred after a short time, and the animals were exposed to the resulting gases and vapours for 1 hour. There were no signs of toxicity or irritation either during the study or in the 2-week observation period (Bayer, 1972).

In a further experiment, 4 g of the mixture was warmed to 120 °C in 40 g of Silicon oil M 100. This caused marked foam-formation, and white flakes were seen in the oil. The animals were exposed to the gases and vapours either once for 4 hours or for 4 hours/day on 5 consecutive days. The observation period was 2 weeks in each case. The results are summarized in Table 1.

Table 1. Inhalation exposure of mice, rats, hamsters and rabbits to a mixture of ADC (90%) and zinc salt (10%)

Concentration in g nominal	analytical	Duration of exposure (hours)	Species	Results
4.0	1.81	4	20 male mice	No effect
			10 male rats	No effect
4.0	1.75	4	5 male hamsters	No effect
			2 male rabbits	No effect
4.0	1.292	5 times 4	20 male mice	No effect
			10 male rats	No effect
4.0	1.59	5 times 4	5 male hamsters	No effect
			2 male rabbits	No effect

As shown in the table the exposed animals were not harmed by the gases and vapours that developed (Bayer, 1972).

Groups of 10 male rats were given a daily dose of Porofor ADC on 5 consecutive days, and then observed for 7 days. The test sample was administered both orally and by intraperitoneal injection. The following doses were administered orally: 0.1, 0.2 and 0.3 g/kg body weight. No deaths occurred. Only the rats in the highest dose group showed transitory symptoms. The following doses were administered by intraperitoneal injection: 0.045, 0.09 and 0.145 g/kg body weight. For 1 rat in the highest dose group, this dose was lethal. The remaining rats in this group and those in the middle dose group showed only transitory symptoms. The lowest dose was without effect. In summary, no cumulative effect of the compound could be detected in this preliminary study (Bayer, 1965).

To determine the effect of inhalation exposure of Porofor ADC, studies involving spraying at a constant or variable concentration were conducted, as were vapour-inhalation studies. For spraying at a constant concentration, 20 male rats were held in a dynamic inhalation apparatus. The observation period was 2 weeks. The compound was sprayed as a mixture with dimethylsulphoxide/polyethylene glycol 400 (1 : 1). Analysis of the air was carried out by spectrophotometry at 262 nm. The results of the study are presented in Table 2.

Table 2. Inhalation exposure of rats on spraying at a constant concentration

Nominal concentration (mg/m³)	Analytical concentration (mg/m³)	Duration of exposure (hours)	Toxicological findings [a]
100	22	4	0/0/20
100	35	5 times 4	4/10/20
25	10	5 times 4	2/5/20
10	4	5 times 4	0/0/20

[a] Number of dead animals/number of animals with symptoms/number of animals exposed

The results showed that, on repeated exposure, substantially smaller quantities of Porofor ADC in the air were tolerated by rats than on single exposure (Bayer, 1965).

To study the inhalation toxicity on spraying at an inconstant concentration, 1 g Porofor ADC in 20 ml of a mixture of dimethylsulphoxide/polyethylene glucol 400 (1 : 1) was nebulized 8 times in 4 hours at regular intervals in a 2 m³ chamber in which 1 rabbit, 2 guinea-pigs, 10 rats and 20 mice were placed (age, origin and weight of the animals not specified). 1 Guinea-pig, 3 rats and all of the mice died during the 8-day observation period, but no specific signs of toxicity were observed. In a comparable study, 0.2 g Porofor ADC formulated in the same way was sprayed into the 2 m³ chamber 8 times in 4 hours, producing an average analytical concentration of 10 mg/m³ air. The animals behaved normally, during the exposure and observation periods, and no signs of irritation were observed (Bayer, 1965).

In a vapour-inhalation study, 1 rabbit, 2 guinea-pigs, 10 rats and 20 mice (age, origin and weight not specified) were put in a

400 l chamber, in which 4 g Porofor ADC was heated to 300–350 °C. At this temperature, decomposition of the compound occurred together with marked mist formation and sublimation. The animals behaved normally during the study and the observation period. During the 4-hour exposure period, the decrease in the amount of compound present was 3.88 g (Bayer, 1965).

In a further study with a comparable protocol, 1 cat, 1 rabbit, 2 guinea-pigs, 10 rats and 10 mice were exposed for 4 hours/day on 5 consecutive days to the vapours that arose when 1 g Porofor ADC was heated to 300 to 350 °C. No signs of irritation or toxicity were seen either during the week of exposure or during the 2-week observation period. Repeated exposure to a Porofor ADC aerosol at higher concentrations led to signs of toxicity in the experimental animals, but the vapour liberated on decomposition of Porofor ADC was harmless (Bayer, 1965).

In a lung function test, groups of 20 male Hartley guinea-pigs (strain Charles River, aged 8–9 weeks) were exposed to the following ADC concentrations for 1 hour: 0, 19, 58 and 97 mg/m^3 air (concentration measured analytically). During the study, the animals were in whole-body plethysmographs attached to the exposure chamber. In order to measure transpulmonary pressure in the animals, polyethylene catheters were implanted in the pleural cavity under anaesthesia 24 hours before exposure began. The lung functions measured included tidal volume, tidal flow, transpulmonary pressure, compliance and total pulmonary resistance. In addition, the respiratory frequency and the minute-volume were calculated in relation to the tidal volume. Lung function measurements were carried out before exposure, at 15, 30, 45 and 60 minutes during the exposure period, and 15 minutes after exposure had ended. After each of these measurement periods, 1 guinea-pig from each group was killed and examined. The remaining guinea-pigs underwent a further lung test 24 hours after the end of exposure. At autopsy, the lungs, nasal cavity, larynx, trachea and tracheobronchial lymph nodes were removed, fixed in formalin, processed into 5 µm-thick sections and stained for histopathological examination. The tissues of the control animals and those of the highest concentration group (97 mg/m^3 air) were microscopically examined. There were no macroscopic or microscopic changes, and compliance and total pulmonary resistance were unchanged after 1 hour of ADC exposure. Significant, concentration-dependent decreases in respiratory volume, respiratory frequency and tidal volume were observed, but these changes were slight; the greatest change measured was in the minute-volume (decreased by 24% in the highest dose group). Overall it was shown that concentrations of ADC up to 97 mg/m^3 air led to slight functional changes, but no associated histopathological changes (Shopp et al., 1987).

The effect of ADC on the thyroid gland of male Sprague-Dawley rats (weight 160 to 180 g) was assessed after repeated oral or intraperitoneal administration. In the oral study, groups of 7 to 12 rats received the following doses in their feed: 0, 10, 50 and 100 g/kg feed for 1 or 4 weeks. In the intraperitoneal study groups of 8 male rats were injected with ADC in propylene glycol at doses of 0.2, 2 and 20 mg/100 g body weight for 1 week. One day before the end of each study, 0.1 µCi ^{125}iodine in isotonic saline was injected intraperitoneally into each rat. At autopsy, the thyroid gland was removed and weighed and the iodine uptake by the thyroid gland after 24 hours was determined. In addition, the plasma hormone level and the iodine-protein binding in the plasma were determined. Reduced iodine uptake by the thyroid gland was found after oral administration at the two higher dose levels. However, a struma-inducing effect could not be established. The protein-bound iodine concentration in the blood tended to be lower in the ADC-treated animals. After intraperitoneal administration, 5/8 rats in the highest dose group died during the first week. Haematuria was observed on macroscopic examination in these animals. The function of the thyroid gland was unaffected (Gafford et al., 1971).

Groups of 5 male and 5 female mice (B6C3F1-strain) were given the substance daily 12 times (weekends not included) in corn oil at the following doses: 625, 1250, 2500, 5000 and 10 000 mg/kg body weight. A further 5 male and 5 female mice served as vehicle controls. During the study a total of 19 mice died (6 at 10 000 mg/kg, 5 at 5 000 mg/kg, 6 at 2 500 mg/kg, and 2 at 1 250 mg/kg). The following symptoms were observed at 1 250 mg/kg and above: ruffled fur, reduced activity, reduced limb reflexes and grasp reflexes, bradypnoea and ptosis. Loose stools and diarrhoea were seen in the animals in all dose groups. The body weight of the mice was not adversely affected in comparison with the controls. Pathological examination revealed crystals in the renal tubuli and the bladder, tubular nephrosis and changes in the renal papilla and kidney pelves in all treated males and in the females given 1250 mg/kg and above (International Research and Development Corporation, 1991).

In a comparable study, Fischer-344 rats (same doses, duration of administration and group size) were given the substance by stomach tube. In total, 10 rats died during the experiment (5 at 10 000 mg/kg, 2 at 5000 mg/kg, 2 at 2500 mg/kg and 1 at 1250 mg/kg). Symptoms observed included brown smeared snouts, body tremors, possible respiratory congestion and wheezing. In the females, symptoms occurred that were also seen in the control group. Body-weight gain was not adversely affected in any of the dose groups. Macroscopic and microscopic examination revealed nephritis and nephrosis as well as tubular nephrosis and nephritis, from 1250 mg/kg in the males and from 2500 mg/kg in the females (International Research and Development Corporation, 1991).

Male and female F344/N rats and B6C3F1 mice (ca. 7 weeks old, 5 animals/group) were exposed to azodicarbonamide aerosols by inhalation at average concentrations of 2, 9.4, 52, 102 or 207 mg/m^3 for 6 hours/day, 12 times over a period of 14 days. No deaths or clinical signs of toxicity were observed in the animals either during or after exposure. At the highest dose level, the body weights were slightly below those of the controls. Only the male rats at the highest dose level showed reduced liver weights compared with the controls. No adverse effects were detected macroscopically or microscopically either in the rats or the mice. Similarly, the methaemoglobin level and the activity of acetylcholinesterase in the blood did not differ between the treated and control groups (Medinsky et al., 1990).

7.3 Skin and mucous membrane effects

Tests on skin and eye irritation in rabbits (white New Zealand, weight 3 to 5.4 kg) were carried out during the initial development of Porofor ADC and Porofor ADC/K. A small cotton wool pad containing ca. 50 mg of the test substance was fixed with an adhesive plaster to the inside of the ear of a rabbit for 24 hours. The treated skin was unchanged when the dressing was removed and during the 7-day observation period. To study mucous membrane irritancy, ca. 50 mg of each test substance was instilled into the conjunctival sacs of 2 rabbits. After instillation of Porofor ADC powder, slight transient reddening of the mucous membranes was observed. This was considered by the authors to be mechanically-induced. The instillation of Porofor ADC/K, on the other hand, led to marked redness and swelling of the mucous membranes. However, no changes to the cornea were evident. The negative results of the skin and eye irritation tests have also been confirmed by other authors (Bayer, 1965, 1972, 1977).

Azodicarbonamide was applied 6 times under occlusive cover to the clipped dorsal skin of rabbits (3 per group) as a 50% aqueous solution or a 50% suspension in acetone. After the second application the rabbits showed slight erythema and scaling, but this was no longer evident after the fourth application (ICI, 1966).

3 Rabbits had 0.05 ml of a 25% aqueous suspension of azodicarbonamide instilled into their left eyes. After 1 hour, 2 rabbits showed slight irritation of the conjunctiva, which was reversible after 24 hours. The fluorescein test was negative in all 3 rabbits (ICI, 1966).

7.4 Sensitization

Groups of male Hartley guinea-pigs inhaled aerosols of azodicarbonamide at concentrations of 0 (control), 51 or 200 mg/m^3, 6 hours/day, 5 days/week for 4 weeks. A proportion of the guinea-pigs (10 animals/concentration) underwent tests for specific sensitization, in which the reciprocal value of the resistance of the airways was measured during an inhalation challenge with azodicarbonamide, before and on the third day after the 4-week exposure. The concentration for the challenge reaction was the same as for the repeated exposure. The remaining guinea-pigs (10 animals/concentration) underwent tests for non-specific respiratory sensitization, in which they inhaled a histamine aerosol before and after the 4-week azodicarbonamide exposure. Skin tests for immune response, body weight determinations and histopathological examination of the respiratory tract and the treated skin were carried out. The 4-week exposure caused neither specific nor non-specific respiratory sensitization. No skin reactions, effects on body weight or histopathological effects were found. These results indicate that azodicarbonamide is not a respiratory sensitizer in guinea-pigs under these conditions (Gerlach et al., 1989).

In an older study, the skin sensitizing potential of azodicarbonamide was tested. 4 Alderley Park guinea-pigs (age ca. 8 weeks, sex unspecified) had 0.1 ml of the test substance (1% in dimethylformamide) applied to the outer side of the ear every day for 3 days. 7 Days after the beginning of the study, 0.2 ml of azodicarbonamide at various concentrations (no further details) was applied to the clipped flanks of the guinea-pigs. The results were evaluated 24 hours later. Azodicarbonamide was not a sensitizer under these conditions (Stevens, 1967).

7.5 Subchronic and chronic toxicity

Groups of 10 male and 10 female Fischer-344 rats (weight not specified) were given ADC in corn oil by stomach tube daily for 13 weeks at the following dose levels: males 100, 500 and 2500 mg/kg; females 200, 1000 and 5000 mg/kg. A group of 10 males and 10 females received the vehicle alone (control group). At the highest dose level, all the males and 2 of the 10 females died during the study. The following symptoms were observed in these animals: reduced activity, dehydration, ruffled fur, bradypnoea, bradycardia, reduced body temperature, irregular heart beat, watering eyes, wheezing, clonic spasms, emaciation, abnormal posture, reduced limb and grasp reflexes and ptosis. The average body weight of the surviving animals in all dose groups was unchanged in comparison with the controls. Macroscopic, microscopic and pathological examinations at the end of the study revealed pyelonephritis, with concretions and crystal deposits in the renal tubuli of the top-dose males and females that died during the treatment. In this study, the no effect level for males was between 500 and 1000 mg/kg, and for females was 1000 mg/kg (International Research and Development Corporation, 1991).

In a comparable study, 10 male and 10 female B6C3F1 mice were given the following doses by stomach tube for 13 weeks (5 days/week): males 78, 156, 312, 625 and 1250 mg/kg body weight; females 156, 312, 625, 1250 and 2500 mg/kg body weight. A further group of 10 male and 10 female mice received the vehicle alone (control group). No deaths occurred during the treatment period. No statistically significant differences in body weight gain were observed. Macroscopic and microscopic examination of the organs revealed no substance-related abnormalities. The absolute and relative weights of various organs were significantly different, in particular the lung weight was increased in various treatment groups. However, because of the absence of associated morphological effects, the organ weight changes were judged by the authors not to be toxicologically significant. Thus, noeffect levels of 1250 mg/kg for males and 2500 mg/kg for females were reported. The authors referred to the range-finding study that had previously been carried out (see Sect. 7.2), in which clear effects on the target organ (kidney) were found (International Research and Development Corporation, 1991).

Groups of 10 male and 10 female F344/N rats (ca. 7 weeks old) inhaled azodicarbonamide in aerosol form for 6 hours/day, 5 days/week, for 13 weeks, at average concentrations of 50, 100 and 204 mg/m^3. No deaths or clinical symptoms were observed during the study. The body weights of the exposed animals were not significantly different from those of the controls at the end of the study. Increased lung weights were found in both males and females at the lowest exposure level. In addition, enlarged bronchial and mediastinal lymph nodes, perivascular aggregation of lymphocytes and the occurrence of inflammatory cells were observed. The authors considered that the cause of this immune response was more likely to be an unknown virus or viral antigen than an antigenic effect of azodicarbonamide, as these effects were not found in the higher exposure groups. There were no effects in any of the other organs. The blood parameters were not affected in any of the exposed animals, nor were there any differences in the four urinary enzymes studied (LDH, AP, β-G, NAG). In the males, the levels of T_3 and T_4 increased with increasing azodicarbonamide concentration, the increase being statistically significant at the highest concentration. In contrast, no effects were observed in the females (Medinsky et al., 1990).

In an analogous study, groups of 10 male and 10 female B6C3F1 mice (ca. 7 weeks old) were used. No treatment-related deaths occurred during this study. The body weights of the males at the middle and high exposure levels and of the females at the high exposure level were significantly lower than those of the controls. The liver weights at the highest exposure level were significantly reduced in both sexes. There were, however, as in the other organs examined, no macroscopic or microscopic findings. Thus, inhalation exposure to azodicarbonamide concentrations of up to 200 mg/m^3 for 13 weeks appeared not to be toxic in rodents (Medinsky et al., 1990).

7.6 Genotoxicity

7.6.1 In vitro

ADC was tested in the Salmonella/microsome assay in the *Salmonella typhimurium* strains TA 100, TA 1535, TA 1537 and TA 98, with and without S9-mix (from Aroclor-induced liver fractions from Sprague-Dawley rats and Syrian hamsters), with pre-incubation for 20 minutes at 37 °C. The following doses were tested: 0, 33.3, 100, 333, 1000, 3333 and 10 000 µg/plate. Appropriate positive controls were also used. Under these test conditions ADC was mutagenic in strains TA 100 and TA 1535 with and without S9-mix (Mortelmans et al., 1986).

In a further study in which a comparable protocol was used, ADC was positive only in strain TA 100 with and without metabolic activation. In contrast, studies in *Escherichia coli*, similarly with and without the addition of S9-mix, were negative (no further details; Hachiya, 1987).

7.6.2 In vivo

ADC was tested in *Drosophila melanogaster* for sex-linked and recessive lethal effects. When injected into male Drosophila flies at doses of 8000 and 10 000 ppm, ADC showed no evidence of point-mutagenic activity (Yoon et al., 1985).

In a micronucleus test in the mouse, ADC had no effect on the frequency of micronucleus formation in erythrocytes after intraperitoneal injection of 20, 40 and 80 mg/kg body weight, formulated in olive oil (abstract of a Japanese report; Hachiya, 1987).

In another standard plate test azodicarbonamide (dissolved in DMSO) was tested at doses of 100–6667 µg/plate in the *Salmonella typhimurium* strains TA 98 and TA 100 with and without metabolic activation (S9-mix from Aroclor 1254 induced rat or hamster liver). In strain TA 98 the substance showed negative results either with or without metabolic activation, whereas the substance was point-mutagenic in strain TA 100 (NCI, 1988).

In a study of 237 chemicals used as food additives, 19 were positive in the Salmonella/microsome test, including azodicarbonamide (no further details; Zeiger and Dunkel, 1991).

Azodicarbonamide was negative in a sister chromatid exchange study. The results of a chromosomal aberration assay were equivocal (NTP, 1985).

In the mouse lymphoma test azodicarbonamide (dissolved in DMSO) showed negative results in L5178Y (TK^+/TK^-) cells without metabolic activation at doses of 73–150 µg/ml. 106–261 µg/ml also showed negative results with metabolic activation (S9-mix from Aroclor 1254 induced rat liver; NCI, 1988).

7.7 Carcinogenicity

In a study in rats in which ADC was added to a special diet prior to baking, the presence of ADC was not incontestably demonstrated. The results are therefore not valid in any assessment of carcinogenic activity (Oser et al., 1965).

7.8 Reproductive toxicity

In parallel with the chronic feeding study described in Sect. 7.7, reproduction and lactation efficiency were also studied. For the reasons given in Sect. 7.7, the results are of little value (Oser et al., 1965).

In a 90-day inhalation study in F344/N rats and B6C3F1 mice (for study design see Sect. 7.5), sperm morphology and vaginal cytology were assessed. In the males, there were no treatment-related effects on the weights of the testes and epididymides or on the number and motility of sperm cells. No abnormal sperm cells were seen. In the females, there were no effects on the regularity or length of the oestrus cycle (Medinsky et al., 1990).

7.9 Effects on the immune system

No information available.

7.10 Neurotoxicity

Acute oral administration of Porofor-505 to mice led to inhibition of pseudo-acetylcholinesterase and acetylcholinesterase activity in the blood and cholinesterase activity in the liver (abstract of a Russian report, no further details; Melinikova and Selikhova, 1965).

7.11 Other effects

No information available.

8. Experience in humans

Workers who were in contact with ADC as a dust-like powder developed irritation of the eyes and the respiratory airways, followed by coughing, shortness of breath and wheezing. On longer contact, delayed asthma occurred. Lung function tests with spirometry revealed that both the forced vital capacity (FVC) and the forced expiratory volume (FEV_1) were slightly to moderately reduced. Repeated contact caused hyperactivity of the respiratory system. The substance was therefore considered to be potentially sensitizing in the respiratory tract by the various authors as an indication of a respiratory sensitization effect (Ferris et al., 1977; Slovak, 1981, 1984; Danan et al., 1984; Valentino and Comai, 1985; Malo et al., 1985; Whitehead et al., 1987; Normand et al., 1989).

In addition to the lung sensitization effect, a few cases of contact allergy have been described in workers after handling ADC (Nava et al., 1983; Bonsall, 1984; Yates and Dixon, 1988).

References

Bayer AG, Insitut für Toxikologie
Toxikologische Untersuchungen mit Porofor ADC, Urazol und Schaumstoffextrakten
Unpublished report of 26.05.1965

Bayer AG, Institut für Toxikologie
Versuchsprodukt KA 9036, Toxikologische Untersuchungen
Unpublished report no. 3372 (1972)

Bayer AG, Institut für Toxikologie
Porofor ADC/K, akute orale Toxizität
Unpublished report of 15.06.1976

Bayer AG, Institut für Toxikologie
Porofor ADC/K, Untersuchungen zur Haut- und Schleimhautverträglichkeit
Unpublished report of 17.01.1977

Bayer AG
Grunddatensatz für Altstoffe über 1000 jato, Diazenedicarboxamide, 31.07.1989

Bechtold, W.E., Medinsky, M.A., Cheng, Y.S., Hobbs, C.H.
Azodicarbonamide; methods for the analysis in tissues of rats and inhalation disposition
Xenobiotica, 19 (9), 1003–1012 (1989)

Bonsall, J.L.
Allergic contact dermatitis to azodicarbonamide
Contact Dermatitis, 10, 42–61 (1984)

Danan, M., Conso, F., Dally, S., Diamant-Berger, O., Desnoyers, G., Gervais, P.
Asthmes professionels dans l'industrie du caoutchouc, a propos de 4 observations
Arch. Mal. Prof., 45, 442–443 (1984)

Eichenhofer, K.W., Roos, E.
Hydrazin
In: Ullmann's Enzyclopädie der technischen Chemie
Vol. 13, p. 95–107
Verlag Chemie, Weinheim (1977)

Ferris, B.G., Peters, J.M., Burgess, W.A., Chery, R.B.
Apparent effect of an azodicarbonamide on the lungs
J. Occup. Med., 19, 425 (1977)

Gafford, F.H., Scharry, P.M., Pittman, J.A,. jr.
Effect of azodicarbonamide (1,1'-azobisformamide) on thyroid function
J. Clin. Endocrinol. Metab., 32, 659–662 (1971)

Gerlach, R.F., Medinsky, M.A., Hobbs, C.H., Bice, D.E., Bechtold, W.E., Cheng, Y.S., Gillett, N.A., Birnbaum, L.S., Mauderly, J.L.
Effect of four-week repeated inhalation exposure to unconjugated azodicarbonamide on specific and non-specific airway sensitivity of the guinea pig
J. Appl. Toxicol., 9 (3), 145–153 (1989)

Hachiya, N.
Evaluation of chemicals genotoxicity by a series of short-term tests
Akita Igata, 14, 269–292 (1987)

ICI (Imperial Chemical Industries Ltd.), Industrial Hygiene Research Laboratories
Toxicological report – azodicarbonamide ("Genitron" AC4)
Unpublished report no. TR/522 (1966)

International Research and Development Corporation
Biological disposition of orally administered azodicarbonamide in male rats
Unpublished report to the National Toxicology Program, report no. 5703–123 (1982 a)

International Research and Development Corporation
Acute inhalation toxicity test with azodicarbonamide in rats
Unpublished report to the National Toxicology Program, report no. 5793–115 (1982 b)

International Research and Development Corporation
Subacute and subchronic oral toxicity of azodicarbonamide in rats and mice
Unpublished report supplied to BG Chemie by U.S. Department of Health and Human Services, February 14, 1991

Malo, J.L., Pineau, L., Cartier, A.
Occupational asthma due to azobisformamide
Clin. Allergy, 15, 261–264 (1985)

Medinsky, M.A., Bechtold, W.E., Birnbaum, L.S., Bond, J.A., Burt, D.G., Cheng, Y.S., Gillett, N.A., Gulati, D.K., Hobbs, C.H., Pickrell, J.A.
Effect of inhaled azodicarbonamide on F344/N rats and B6C3F1 mice with 2-week and 13-week inhalation exposures
Fundam. Appl. Toxicol., 15, 308–319 (1990)

Melinikova, E.A., Selikhova, L.N.
The blood and liver cholinesterase activity in poisoning with Porofor-505 and Celogen
Farmakol. Toksikol., 28, 742–744 (1965)

Mewhinney, J.A., Ayrs, P.H., Bechtold, W.E., Dutcher, J.S., Cheng, Y.S., Bond, J.A., Medinsky, M.A., Henderson, R.F., Birnbaum, L.S.
The fate of inhaled azodicarbonamide in rats
Fundam. Appl. Toxicol., 8, 372–381 (1987)

Mortelmans, K., Haworth, S., Lawlor, T., Speck, W., Tainer, B., Zeiger, E.
Salmonella mutagenicity tests: II. Results from the testing of 270 chemicals
Environ. Mutagen., 8, 1–119 (1986)

National Polychemicals, Inc.
Technical Bulletin No. OJE-02-0766
Cited in: Mewhinney et al. (1987)

Nava, C., Beretta, F., Elena, A., Glizzi, A., Pattarin, R., Villa, L.
Le dermatiti allergiche de miglioranti e da altri additive par farmie
Med. Lav., 74, 376–379 (1983)

NCI (National Cancer Institute)
Short-term test program sponsored by the Division of Cancer Etiology (1988)

Normand, J.C., Grange, F., Hernandez, C., Ganay, A., Davezies, R.P., Bergeret, A., Prost, G.
Occupational asthma after exposure to azodicarbonamide: report of four cases
Br. J. Ind. Med., 46, 60–62 (1989)

NTP (National Toxicology Program)
Annual plan for fiscal year 1985
U.S. Department of Health and Human Services (1985)

Oser (1965)
Cited in: BIBRA toxicity profile – azodicarbonamide (1990)

Oser, B.L., Oser, M., Morgareidge, K., Sternberg, S.S.
Studies of the safety of azodicarbonamide as a flour-maturing agent

Toxicol. Appl. Pharmacol., 7, 445–472 (1965)

Shopp, G.M., Cheng, Y.S., Gillett, N.A., Bechtold, W.E., Medinsky, M.A., Hobbs, C.H., Birnbaum, L.S., Mauderly, J.L.
Acute inhalation exposure of azodicarbonamide in the guinea pig
Am. Ind. Hyg. Assoc. J., 48, 127–132 (1987)

Slovak, A.J.M.
Occupational asthma caused by a plastics blowing agent, azodicarbonamide
Thorax, 36, 906–909 (1981)

Slovak, A.J.M.
Occupational hazards of polyethylene and polypropylene processing
Iss. Industrial hazards of plastics and synthetic elastomers
Prog. Clin. Biol. Res., 141, 313–317 (1984)

Stevens, M.A.
Use of the albino guinea-pig to detect the skin-sensitizing ability of chemicals
Br. J. Ind. Med., 24, 189–202 (1967)

Uniroyal Chemical
Celogen AZ, product safety data sheet
Cited in: Whitehead et al. (1987)

Valentino, M., Comai, M.
Asma professionale da azodicarbonamide: caso clinico
G. Ital. Med. Lev., 7, 97–99 (1985)

Whitehead, L.W., Robins, T.G., Fine, L.J., Hansen, D.J.
Respiratory symptoms associated with the use of azodicarbonamide foaming agent in a plastics injection molding facility
Am. J. Ind. Med., 11, 83–92 (1987)

Yates, V.M., Dixon, J.E.
Contact dermatitis from azodicarbonamide in earplugs
Contact Dermatitis, 19, 155–156 (1988)

Yoon, J.S., Mason, J.M., Valencia, R., Woodruff, R.C., Zimmering, S.
Chemical mutagenesis testing in drosophila. IV. Results of 45 coded compounds
Tested for the National Toxicology Program
Environ. Mutagen., 7, 349–367 (1985)

Zeiger, E., Dunkel, V.C.
Mutagenicity of chemicals added to foods
In: Hayatsu, H. (ed.)
Mutagens in food: detection and prevention, p. 51–56
CRC Press, Boca Raton, Florida, USA (1991)

Isobutylidenediurea

1. Summary and assessment

12 Hours after feeding of a semi-synthetic diet containing 60 g ^{15}N-isobutylidenediurea/day to sheep, 17.3 mg of the substance (≙2.2% of the last administration) have been determined in 100 mg blood plasma, and only a minimal amount of isobutylidenediurea (max. 1%) and ^{15}N-isobutylidenediurea (max. 0.2%) have been found in the urine. Resorption of the substance through the rumen have been shown to occur with one of the sheep with a ligature at the abomasum entry, and elimination of the substance has been shown to occur only to minimal extent through the kidney.

The acute oral toxicity (LD_{50}) of isobutylidenediurea in the rat is >10 000 mg/kg body weight, and in the mouse following intraperitoneal administration is ca. 500 mg/kg body weight. Signs of toxicity include tachypnoea, dyspnoea, apathy and lying on the stomach. In young cattle and sheep a single administration of 5 g isobutylidenediurea/kg body weight with the drinking water have been caused refusal of feed and drinking water as well as an increased frequency of respiration, miction and defecation, and fibrillation of the musculature of the extremeties. After 7.5 g/kg, 1 of 2 sheep has died and 10 g/kg body weight have caused the death of both of 2 sheep investigated.

Rats are able to tolerate inhalation exposure to an atmosphere enriched with isobutylidenediurea at room temperature for 8 hours without signs of toxicity; pathological effects are not observed at autopsy.

Dermal application of isobutylidenediurea to rabbits for 20 hours does not induce skin irritation. The chemical is not irritant to rabbit eyes.

Isobutylidenediurea has shown no point-mutagenic activity or bacteriotoxicity in the Ames test either with or without metabolic activation (S9-mix) at doses of up to 5000 µg/plate.

In an incompletely-documented carcinogenicity study that does not comply with current requirements, isobutylidenediurea showed no indications of carcinogenic potential in mice or rats. (According to the authors there has also been no "statistically significant" effect on the incidence, time of appearance or location of the observed tumours.)

After oral administration of 100–1000 mg isobutylidenediurea/kg body weight by gavage from day 6–15 post coitum to pregnant rats the substance has neither shown maternal- or embryo-/fetotoxicity nor teratogenicity.

2. Name of substance

2.1 Usual name — Isobutylidenediurea

2.2 IUPAC-name — 1,1-Isobutylidenediurea

2.3 CAS-No. — 6104-30-9

3. Synonyms, common and trade names

Isobutyldiurea
Isobutylenediurea
1,1-Isobutylidenediurea
1,1-Isobutylidene-bis-urea
N,N''-(2-Methylpropylidene)-bis-urea
N,N''-(Isobutyliden)diharnstoff
Urea, 1,1'-isobutylidenedi-
Urea, N,N''-(2-methyl-propylidene)bis-
Diureidoisobutane
1,1-Diureidoisobutane
IBDH
Isodur
IBDU
Isobutylidendiharnstoff
Isobutyldiharnstoff
Isobutylendiharnstoff
1,1-Isobutylidendiharnstoff

4. Structural and molecular formulae

4.1 Structural formula

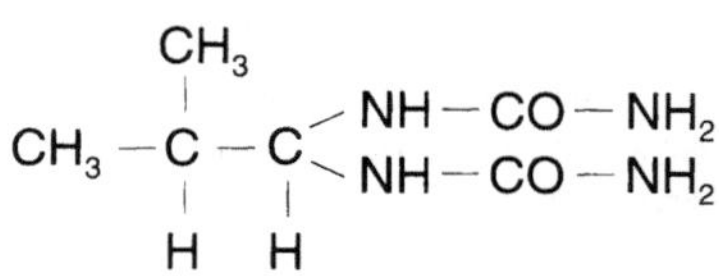

4.2 Molecular formula — $C_6H_{14}N_4O_2$

5. Physical and chemical properties

5.1	Molecular mass, g/mol	174.24
5.2	Melting point, °C	195–205 (depending on water and urea content) (BASF, 1987)
5.3	Boiling point, °C	No information available
	Decomposition temperature, °C	Pure product, from 100; Technical grade product, from 140; Formation of flammable vapours possible (BASF, 1987)
5.4	Vapour pressure, hPa	No information available
5.5	Density, g/cm^3	No information available
5.6	Solubility in water	Pure product: 2 g/l (at 20 °C) (BASF, 1987; 1988 a ca. 99 g/l (at 100 °C) (BASF, 1988 a) Technical grade product: 10.5–14 g/l (at 20 °C) (BASF, 1987)
5.7	Solubility in organic solvents	Completely soluble in DMSO (BASF, 1988 b)
5.8	Solubility in fat	No information available
5.9	pH-value	4–7 (at 100 g/l water) (BASF, 1988a)
5.10	Conversion factor	1 ppm ≙ 7.11 mg/m^3 1 mg/m^3 ≙ 0.14 ppm (at 25 °C and 1013 hPa)

6. Uses

Long-term nitrogenous fertilizer (depot fertilizer) for agriculture and horticulture (Rother, 1975; Neumüller, 1983).

7. Experimental results

7.1 Toxicokinetics and metabolism

4 Mutton Merino wethers averaging 47.6 kg body weight were fed a semi-synthetic diet containing, as the only nitrogen source, ^{15}N-labelled isobutylidenediurea (60 g/day). $2^{1}/_{2}$, $7^{1}/_{4}$, 12 and 24 hours after isotope intake the blood plasma, the urine and the liver were analysed for ^{15}N-content, the whole blood plasma for NH_3, urea and isobutylidenediurea, and the urine for these 3 compounds in labelled and unlabelled form, respectively. From another wether provided with a ligature at the abomasum entry 24 hours after isotope intake the same analyses were taken. In all animals an almost equal urea level ranging from 18 to 30 mg/100 ml was detected in the whole blood plasma. Isobutylidenediurea was only detectable in the animal examined 12 hours after the start of the experiment and amounted to 17.3 mg/100 ml blood plasma (≙ 2.2% of the last isobutylidenediurea-intake), and in the ligature-wether in wich the amount was 18 mg/100 ml blood plasma, so that a resorption of the substance through the rumen-wall has to be assumed. In the urine, only slight amounts of isobutylidenediurea (max. 1% of intake) and ^{15}N-isobutylidenediurea (max. 0.2% of the intake) were retrieved. These findings were regarded as evidence, of the possibility of isobutylidenediurea being absorbed through the rumen wall, as well as of the elimination of the substance through the kidney to a very low extent only (Bergner et al., 1978).

7.2 Acute and subacute toxicity

An LD_{50} value of >10 000 mg/kg body weight was reported for rats after a single oral dose of isobutylidenediurea (no details of strain, sex or number of animals). The observed symptoms included tachypnoea. Autopsy revealed no treatment-related pathological effects (BASF, 1967).

In the mouse, an LD_{50} value of ca. 500 mg/kg body weight was determined after a single intraperitoneal injection (no information on strain, sex, or number of animals). The toxicity was characterized by dyspnoea, apathy and lying on the abdomen. Autopsy revealed pale-coloured kidneys and congestion of the kidney medulla (BASF, 1967).

To study the acute toxicity of isobutylidenediurea on inhalation exposure, 12 rats (strain and sex not specified) were exposed for

8 hours at 20 °C to a stream of air which was either saturated with the volatile fraction of isobutylidenediurea or enriched with dust from moderate dust generation. The air was passed through a ca. 5 cm deep layer of isobutylidenediurea (200 l/hour, inhalation hazard test) in order to enrich or saturate it with the chemical. No signs of toxicity were observed in any of the 12 rats during exposure. No pathological effects were found at autopsy after a 7-day observation period (BASF, 1967).

For the investigation of the acute oral toxicity of isobutylidenediurea in ruminants 6 Simmental-bullocks (150–200 kg body weight) received 3 or 5 g/kg body weight, and 12 sheep ("halbfeinwollige Rasse", 35–40 kg body weight) received 5, 7.5 or 10 g/kg body weight in their drinking water. The animals were observed clinically and their blood was examined before and 1, 3, 6 and 24 hours after administration of the substance and then daily for another 7 days. The no effect level in the bullocks was 3 g/kg body weight. 5 g/kg caused refusal of feed and drinking water as well as an increased frequency of respiration, miction and defecation, and fibrillation of the musculature of the extremeties, which was reversible after 3 to 5 hours. The blood examination revealed a 8.6-fold increase of the ammonia nitrogen, a 2-fold increase of the glutamine nitrogen and a 1.7-fold increase of the ketone-body nitrogen. The urea-concentration increased from 19.7 to 71.2 mg % from day 1 to 3, and the rest nitrogen from 40.5 to 72 mg %. In the sheep the same clinical symptoms were observed for 6 hours after administration of 5 g/kg. 7.5 g/kg, however, caused death in 1 of the 2 animals $1^1/_4$ hours after administration of isobutylidenediurea. The prelethal symptoms 30 minutes after administration of the substance were tremor of the whole body, rapid and noisy breathing, tonic-clonic, then tonic cramps, coma, fluid from the nose and mouth, tympany and atony of the abomasum, increased heart frequency. In the second animal investigated the clinical symptoms were reversible after 1 day. The administration of 10 g/kg body weight caused the same symptoms of intoxication and death after 1.5 hours. The post-mortem examination showed a severe distension of the abomasum, foamy fluid in the oesophagus, trachea and bronchii and signs of haemorrhagic inflammation in the rumen and small intestine. Spleen and liver were hyperaemic, and in the kidney spotted bleedings were found (Gladenko et al., 1980).

7.3 Skin and mucous membrane effects

The skin irritancy of isobutylidenediurea was studied in a patch test on the clipped dorsal skin of white rabbits. A cotton patch (ca. 2.5 cm×2.5 cm) loaded with a 50% aqueous preparation (paste) of isobutylidenediurea was applied once to the skin for 1, 5 or 15 minutes or for 20 hours (no further details). No irritant effects were seen after 24 hours or 8 days when the results were evaluated (BASF, 1967).

To test for eye irritancy, ca. 50 mg neat powdered isobutylidenediurea was placed in the conjunctival sac of rabbits' eyes. For comparison, talc was introduced into the conjunctival sac of the control eyes (number of animals not specified). In both the control and test eyes, marked reddening of the conjunctiva was seen 1 hour after application. This decreased within 24 hours and was no longer detectable after 8 days. The chemical was thus judged to be non-irritating (BASF, 1967).

7.4 Sensitization

No information available.

7.5 Subchronic and chronic toxicity

No information available.

7.6 Genotoxicity

7.6.1 In vitro

The ability of isobutylidenediurea to induce point mutations was studied in the Ames test in the *Salmonella typhimurium* strains TA 98, TA 100, TA 1535 and TA 1537. Technical grade isobutylidenediurea (31% N, purity 92%) was used in a standard plate test and a pre-incubation test at 5 concentrations of from 20 to 5000 µg/plate, with DMSO as the solvent. Both experiments were carried out with and without metabolic activation by S9-mix (containing microsomal enzymes from Aroclor 1254-induced rat liver). After incubation for 48 hours at 37 °C in each case, isobutylidenediurea showed neither a cytotoxic effect nor point-mutagenic activity (BASF, 1988 b).

7.6.2 In vivo

No information available.

7.7 Carcinogenicity

Male and female mice (weight ca. 18 g each, strain not specified) were given an aqueous emulsion of isobutylidenediurea by stomach tube at a dose of 1600 mg/kg body weight/day, twice weekly for 20 weeks. This corresponded to the maximum tolerated dose (no further details). The total observation period was 92 weeks, after which the animals were examined macroscopically. Tissues that showed pathological changes and any tumour-bearing organs were studied histopathologically. In the 11th month of the study the dose group comprised 108 males and 133 females, the control group 76 males and 91 females (no further details of experimental design). The tumour incidence after 92 weeks is presented in Table 1.

Table 1. Effect of isobutylidenediurea on the tumour incidence in mice

	Sex	Number of mice/group [a]	Number (n) of tumour-bearing animals [b]	Incidence (%) of tumour-bearing animals [b]
Control	Female	91	27	29.6
	Male	76	13	17.2
	Total	167	40	23.4
Isobutylidene-diurea [c]	Female	133	32	24.1
	Male	108	22	20.1
	Total	241	54	22.4

[a] In study month 11 (44 weeks); no tumours were present in animals that had previously died or been killed (no further details)
[b] After 92 weeks
[c] 1600 mg/kg body weight orally, twice weekly for 20 weeks

The manifestation time for tumours was, in both the control and the treatment groups, "essentially" between 13 and 15 months after the beginning of treatment (no further details). The sites of the tumours are shown in Table 2.

Table 2. Effect of isobutylidenediurea on tumour location in mice

	Incidence (%) of mice with tumours	
	Control	Isobutylidenediurea
Lung adenomas	11.3	9.5
Lung adenocarcinomas	1.7	1.2
Mesotheliomas	1.1	1.6
Mammary tumours (Adenofibromas, epidermal cancer)	4.7	6.2
Leukoses	4.7	2.4
Fibromas	–	0.8

"Statistically" (no further details) isobutylidenediurea had no "substantial" influence on the incidence (Table 1), the manifestation time or the location (Table 2) of the observed tumours in mice under the experimental conditions described (Didenko et al., 1979).

Male and female rats (weighing ca. 80 g each, strain not specified) were given an aqueous emulsion of isobutylidenediurea by stomach tube at a dose of 800 mg/kg body weight/day, twice weekly for 20 weeks. This corresponded to the maximum tolerated dose (no further details). The total observation period was 92 weeks, after which the animals were examined macroscopically. Tissues that showed pathological changes and any tumour-bearing organs were studied histopathologically. The tumour incidence at the end of the study period in the 82 treated rats and the 63 control rats is shown in Table 3 (no further details of experimental protocol). No sex-specific differences were seen in the distribution of the tumours. No details were provided of the location of the tumours. The tumour incidence was 7.9% in the control group and 7.3% in the treated group (Table 3). Thus there were no indications that isobutylidenediurea was carcinogenic in this study (Didenko et al., 1979).

Under the study conditions described, there was no evidence that isobutylidenediurea was carcinogenic to rats or mice. However, no conclusion on its carcinogenic activity can be drawn from these studies as they are not reported in full. For example, there is no detailed information on the choice of dose or on the systemic ef-

Table 3. Effect of isobutylidenediurea on the tumour incidence in rats.

	Number of rats/group [a]	Number (n) of rats with tumours [b]	Incidence (%) of rats with tumours [b]
Control	63	5	7.9
Isobutylidenediurea [c]	82	6	7.3

[a,b] At the end of the observation period after 92 weeks
[c] 800 mg/kg body weight orally, twice weekly for 20 weeks

fects (body weight gain, feed consumption, mortality). Furthermore, the substance was only administered twice weekly for 20 weeks (instead of daily administration for life).

7.8 Reproductive toxicity

In a prenatal toxicity study according to OECD-guideline 414, 22 to 24 pregnant Wistar-rats/group received 100, 400 or 1000 mg/kg body weight isobutylidenediurea per gavage from day 6–15 post coitum. The control group was given the vehicle only (0.5% aqueous carboxymethyl cellulose solution) in the same standard dose volume of 10 ml/kg body weight. There were no signs of maternal and embryo/fetotoxicity or teratogenicity up to and including a dose of 1000 mg/kg body weight/day, a dose sufficient for a limit-test. In a pre-study for testing of maternal toxicity even 1200 mg/kg body weight caused no maternal toxicity when body weight gain, hematological, clinico-chemical and urinalytical parameters as well as organ weights were considered. The no effect level in the maternal and fetal organism was 1000 mg/kg body weight (BASF, 1993).

7.9 Effects on the immune system

No information available.

7.10 Neurotoxicity

No information available.

7.11 Other effects

No information available.

8. Experience in humans

No information available.

References

BASF AG, Gewerbehygienisch-Pharmakologisches Institut
Ergebnis der gewerbetoxikologischen Prüfung (XVII 341)
Unpublished report (1967)

BASF AG
Isobutylidendiharnstoff
Grunddatensatz für Großstoffe (1987)

BASF AG
Isobutylidendiharnstoff
DIN-Sicherheitsdatenblatt (1988 a)

BASF AG, Abteilung Toxikologie
Report on the study of Isobutylidendiharnstoff techn. (31% N) in the Ames test
Unpublished report (1988 b)

BASF AG, Abteilung Toxikologie
Report, study of the prenatal toxicity of isobutylidene-di-urea in rats after oral administration (gavage), project no. 30R0727/90116
Unpublished report (1993)
On behalf of BG-Chemie, Heidelberg

Bergner, H., Görsch, R., Adam, K.
Dynamik von ^{15}N-Isobutylidendiharnstoff (IBDH) bei Schafen. 2. Mitteilung, Untersuchungen zum IBDH-Umsatz im intermediären Stoffwechsel
Arch. Tierernährung, 28 (8), 575–583 (1978)

Brunnmüller, F.
Aminoplaste
In: Ullmanns Encyklopädie der technischen Chemie
Vol. 7, p. 403–424
Verlag Chemie, Weinheim (1974)

Didenko, G.G., Gupalovic, T.D., Petrovskaja, O.G.
Carcinogenic properties of iso butylidene di urea
Vopr. Pitan., 37 (2), 72–73 (1979)

Gladenko, I.N., Zajceva, L.D., Śevcova, G.N.
Toxizität des Isobutylidendiharnstoffs (German translation from the Russian)
Veterinarija (Moscow), 10, 60–61 (1980)

Neumüller, O.A. (ed.)
Römpps Chemie-Lexikon
Vol. 3, p. 1949
Franckh'sche Verlagshandlung Stuttgart (1983)

Rother, E.
Depotdünger
In: Ullmanns Encyclopädie der technischen Chemie
Vol. 10, p. 232–238
Verlag Chemie, Weinheim (1975)

Hydroxylamine and its salts

1. Summary and assessment

Hydroxylamine is formed as a metabolite during cellular metabolism. Hydroxylamine reductase can be detected in a number of mammals, and its activity appears to be age-dependent. Metabolic oxidation of hydroxylamine to nitrate has also been described.

Studies on the acute toxicity of hydroxylamine and its salts have given variable results. Based on the majority of the oral LD_{50} values for the rat, the chemical is of moderate toxicity (LD_{50} rat 192 to about 600 mg/kg). This is also the case for the dermal LD_{50} values in the rabbit (1500 to 2000 mg/kg). Effects on the blood have, however, been observed in rabbits after a single dermal application (occlusive) of 10 mg/kg body weight. There is no acute inhalation hazard from hydroxylamine hydrochloride or hydroxylamine sulphate. Likewise, the inhalation of vapour from saturated aqueous solutions of hydroxylamine hydrochloride or hydroxylamine sulphate for 1 hour (no details of concentration) is tolerated by rats without any signs of toxicity.

The target organ for hydroxylamine and its salts in the mouse, rat, rabbit and cat is primarily the erythrocyte, resulting in methaemoglobin formation, Heinz body formation, and secondary anaemia with reticulocytosis and spleen enlargement. Lethal and sub-lethal doses cause central excitation and convulsions.

Repeated subcutaneous administration of 5 mg/kg to rabbits for 1 week leads to severe toxic effects with methaemoglobin levels of up to 60%, and death. A dose of 1 mg/kg over the same time period causes moderate anaemia but not methaemoglubinaemia.

The skin irritancy of hydroxylamine and its salts is only slight, but eye irritancy is severe and only slightly reversible, or irreversilbe, within 1 week.

According to the available data, hydroxylamine and its salts have a sensitizing effect on dermal application and on injection.

After administration of hydroxylamine sulphate in the drinking water of rats for 3 months at concentrations of 250 and 50 ppm (equivalent to ca. 21 and 5 mg/kg body weight/day, respectively), the animals have developed anaemia (dose-related), with corresponding haematological changes and haemosiderosis. No substance-related effects have been found at a concentration of 10 ppm (equivalent to about 1 mg/kg body weight/day).

Atrophy of the thyroid gland and a 4- to 5-fold enlargement of the spleen have been described in rats given milk containing hydroxylamine hydrochloride for 6 months at doses of 333 to 380 mg/kg body weight. After intravenous injection of 2.5 mg hydroxylamine hydrochloride/kg for a period of up to 19 months, rabbits show methaemoglobin formation, anaemia and reticulocytosis. Administration of hydroxylamine sulphate to mice in their drinking water for 12 weeks (1640 and 3280 mg/l) resulted in anaemia, leucocytosis and spleen enlargement. These effects are reversible after 8 and 18 weeks, respectively.

Numerous studies have investigated the genotoxic potential of hydroxylamine and its salts, but most have not been carried out according to the methods usual today, and the results are not consistent. Both positive and negative results have been obtained in in vitro tests for the induction of point mutations, but two Salmonella/microsome tests which have been carried out independently of each other gave negative results both with and without metabolic activation. Tests for clastogenic activity and DNA exchange have proved positive, almost without exception. Cell transformation tests have again given both positive and negative results. In vivo, most studies on *Drosophila melanogaster* have been positive, while in the dominant lethal assay in mice hydroxylamine sulphate was without effect. The results of these studies indicate that hydroxylamine and its salts have genotoxic potential. They appear to act on the chromosomes rather than inducing point-mutations in the genes.

There have been no specific carcinogenicity studies on hydroxylamine and its salts that comply with current requirements. However, in older studies, no treatment-related tumours are found in mice given 1–2% hydroxylamine hydrochloride in their feed for their whole lives. Compared with the controls the treated mice have had a prolonged survival time, but this cannot be confirmed in a more recent study. In another study, the administration of hydroxylamine sulphate in the drinking water of mice at levels of 1640 and 3280 mg/l for 52 weeks suppresses the formation of spontaneous mammary tumours. In this case hormonal factors appear to be involved. In more recent studies, increased incidences of lymphomas and haemangiosarcomas of the spleen and lymph nodes are reported.

Embryotoxic or teratogenic effects of hydroxylamine or hydroxylamine hydrochloride are mostly only observed under exceptional study conditions (injection into the embryonal body cavity or intravenous injection into rabbits, injection into the air space of hens' eggs). Intraperitoneal injection causes no malformations in the offspring of rats, even at doses that are toxic to the dams. Calves from cows treated with hydroxylamine hydrochloride in the feed from the 3rd month of pregnancy until birth show no malfomations.

The motor excitability and convulsions which occur after lethal or sub-lethal doses are likely to be a result of hypoxaemia due to methaemoglobin formation, and are therefore secondary.

Other effects of hydroxylamine and its salts described in the literature are of lesser significance in the overall toxicological assessment. Intravenous injection of 2 to 8 mg hydroxylamine/kg in cats and dogs leads to a decrease in blood pressure, which is more slowly reversed than that induced by sodium nitrite.

In hydroxylamine workers, methaemoglobin concentrations of up to 25% have occurred at the end of the working day. The accidental ingestion of "two gulps" of a hydroxylamine solution has caused nausea, vomiting, cyanosis and haemolytic anaemia that was only slowly reversible. Skin irritation due to hydroxylamine hydrochloride has been described in humans. Diagnostic tests for skin allergy to hydroxylamine hydrochloride and hydroxylamine sulphate have proved positive.

A micronucleus test and an oral teratogenicity study in rats with hydrooxylamine sulphate are beeing performed on behalf of the BG Chemie. A carcinogenicity staudy will be contucted by the BASF A6 according to committee of the BG Chemie.

2. Name of substance

2.1	Usual name	Hydroxylamine
2.2	IUPAC-name	Hydroxylamine
2.3	CAS-No.	7803-49-8 (Hydroxylamine) 5470-11-1 (Hydroxlamine hydrochloride) 10039-54-0 (Hydroxylamine sulphate)

10046-00-1 (Hydroxlamine bisulphate)

3. Synonyms, common and trade names

Hydroxylammonia
Oxyammoniak
Hydroxylammoniak
Hydroxylamin

4. Structural and molecular formulae

4.1	Structural formula	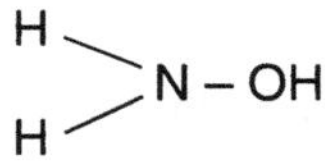
4.2	Molecular formula	H_3NO

5. Physical and chemical properties

5.1	Molecular mass, g/mol	33.04 (Hydroxylamine) 69.50 (-Hydrochloride) 164.14 (-Sulphate) 131.11 (-Bisulphate)
5.2	Melting point, °C	33.1 (Hydroxylamine) 57 (-Bisulphate) 151.0 (-Hydrochloride) 170.0 (-Sulphate) (Malle, 1977)
5.3	Boiling point, °C	56–57 (at 29 hPa; Hydroxylamine) (Malle, 1977)
5.4	Vapour pressure, hPa	0.36 (at 0 °C; Hydroxylamine) 7.1 (at 32 °C; Hydroxlamine) (Malle, 1977)

5.5	Density, g/cm^3	1.216 (Hydroxylamine) 1.67 (-Hydrochloride) 1.88 (-Sulphate) (Malle, 1977)
5.6	Solubility in water	Hydroxylamine and its hydrochloride, bisulphate and sulphate salts are very readily soluble (Malle, 1977)
5.7	Solubility in organic solvents	Hydroxlamine is very radily soluble in methanol and ethanol, very slightly soluble in benzene, chloroform, ethyl acetate and ether (Malle, 1977)
5.8	Solubility in fat	No information available
5.9	pH-value	Alkaline (Gross and Smith, 1985) 3.4 (-Hydrochloride, 0.1 M solution) 3.7 (-Sulphate, 0.1 M solution) 1.6 (-Bisulphate, 0.1 M solution) (Malle, 1977)
5.10	Conversion factor	Hydroxylamine: 1 ppm ≙ 1.35 mg/m^3 1 mg/m^3 ≙ 0.74 ppm Hydroxylamine hydrochloride: 1 ppm ≙ 2.84 mg/m^3 1 mg/m^3 ≙ 0.35 ppm Hydroxylamine sulphate: 1 ppm ≙ 6.70 mg/m^3 1 mg/m^3 ≙ 0.15 ppm Hydroxlamine bisulphate: 1 ppm ≙ 5.35 mg/m^3 1 mg/m^3 ≙ 0.19 ppm (at 25 °C and 1013 hPa)

6. Uses

To manufacture oximes and lactams, particularly caprolactam; polymerisation regulator in butadiene-styrene mixtures; reducing agent in photographic developing solutions; antioxidant; analytical reagent; colour adjuster in the dying of acrylonitrile fibres and the printing of wool; intermediate in the manfuacture of pesticides and pharmaceuticals (Malle, 1977).

7. Experimental results

7.1 Toxicokinetics and metabolism

Hydroxylamine is formed as a metabolite during normal metabolism. It arises from the enzymatic reduction of nitrate and nitrite, or the oxidation of ammonia. These reactions represent intracellular detoxification reactions (Budovskii, 1976; Gross and Smith, 1985).

In a comparative study of extracts of liver mitochondria from various animal species, it was shown that hydroxylamine hydrochloride is reduced to ammonia by means of mitochondrial membrane-bound hydroxylamine reductase in the presence of NADH. The mitochondrial activity varies markedly between particular species and individuals, with mice displaying the highest enzyme activity. In rats the concentration of the enzyme rises with increasing age, particularly in the females. Hydroxylamine reductase is also found in the kidneys of rats and cats, but not however in the brain or the serum. Mitochondria from bovine hearts also show no enzyme activity (Bernheim, 1972).

Hydroxylamine-reducing enzymes have also been found in Bacillus subtilis (Klausmeier and Bard, 1954), Escherichia coli and Clostridium perfringens (Woods, 1983), Pseudomonas aeruginosa (Amarger and Alexander, 1968) and Neurospora crassa (Siegel et al., 1965).

The metabolic oxidation of hydroxylamine hydrochloride in rats has been described. 4 Sprague-Dawley rats (194 $\pm$ 7 g) were given a single dose of 20 µmol ^{15}N-hydroxylamine hydrochloride (equivalent to about 7 mg/kg body weight) by stomach tube. The total elimination of ^{15}N-nitrate in the urine was 4.7% of the administered radioactivity. The largest proportion was found on day 4. The nitrate excretion was not increased after induction with 500 mg Arochlor 1254/kg body weight (Saul and Archer, 1983/84).

7.2 Acute and subacute toxicity

The first observations on the toxic effects of hydroxylamine and hydroxylamine hydrochloride on the blood were published over 100 years ago, but will not be discussed further here because they are mainly of historical interest (Raimondi and Bertoni, 1882; Binz, 1888; Lewin, 1889).

Hydroxylamine

An LD_{50} value of 60 mg/kg was determined in mice after intraperitoneal injection (Lewis and Tatken, 1982).

For rats, the subcutaneous LD_{50} was 29 mg/kg and the intraperitoneal LD_{50} 59 mg/kg (Lewis and Tatken, 1982).

After oral administration to dogs, an LD_{50} value of 200 mg/kg was determined (information on signs of toxicity, possible vomiting in dogs and findings at autopsy was not provided; Oil and Hazardous Materials Assistance Data System, 1978).

Hydroxylamine hydrochloride

An LD_{50} of 1.83 mmol/kg body weight ($\triangleq$ 127 mg/kg body weight) was found in female CD1 mice after intraperitoneal injection of the hydrochloride. Within 10 minutes of the administration of 97 to 153 mg/kg body weight, the methaemoglobin content increased to a maximum of 13 to 14%, but rapidly returned to normal (within 60 minutes). Death of the mice after high doses similarly occurred within 10 minutes (Smith and Layne, 1969).

After intraperitoneal injection of mice with 76 mg/kg, the maximum methaemoglobin content of ca. 35% was reached after 7 minutes. In this study, death of the mice also occurred within the time in which the highest methaemoglobin content was reached (7 to 15 minutes; Kruszyna et al., 1982).

Similar results were obtained in mice after intraperitoneal injection of 100 mg/kg (OTS, 1984; Gross and Smith, 1985).

In further studies an intraperitoneal LD_{50} of ca. 180 mg/kg was determined in mice (BASF, 1975).

Oral LD_{50} values reported for mice include 400 mg/kg (OTS, 1984; Gross and Smith, 1985) and 408 mg/kg for male and 419 mg/kg for female mice (Riemann, 1950).

After oral administration to rats, the LD_{50} was 192 ($\pm$ 52) mg/kg (OTS, 1984; Gross and Smith, 1985).

In another study an oral LD_{50} value of ca. 600 mg/kg was reported for Sprague-Dawley rats. Strong convulsions were observed before death. All of the rats had damage to the spleen (BASF, 1975).

In dogs, the reported subcutaneous LD_{50} was 70 mg/kg (OTS, 1984; Gross and Smith, 1985).

In rabbits, a single subcutaneous injection of 5 mg hydroxylamine hydrochloride/kg led to methaemoglobin levels of up to 18% after 1 hour. Repeated administration of this dose for 1 week caused a bad general condition with marked diarrhoea and weight loss in the rabbits and 5 of the 7 rabbits died. The methaemoglobin level in the blood reached up to 60%. After a single subcutaneous injection of 20 mg/kg in rabbits, the methaemoglobin content after 1 hour was 23 to 26% and after 24 hours was 16 to 22%. After the administration of 30 mg sodium nitrite/kg, methaemoglobin levels of 21 to 26% were measured but, in contrast to hydroxylamine hydrochloride, the values were within the normal range again after 24 hours. The subacute daily administration of 1 mg hydroxylamine hydrochloride/kg body weight to rabbits over 1 week (according to the abstract of this Russian paper, the duration of treatment was 1 month) led to moderate anaemia (no further details; Arnoldova and Speransky, 1963).

In an inhalation hazard test at 20 °C, inhalation of the volatile portion of hydroxylamine hydrochloride for 8 hours was tolerated without signs of toxicity by 12 Sprague-Dawley rats (BASF, 1975).

No deaths or toxic effects occurred in rats on inhalation of the mist from a saturated aqueous solution of hydroxylamine hydrochloride for 1 hour (no further details; OTS, 1984).

Hydroxylamine sulphate

In the mouse, an LD_{50} value of ca. 1000 mg/kg was determined after oral administration (Litton Bionetics Inc., 1979/1980).

An oral LD_{50} value of 545 (± 41) mg/kg was reported for rats, and at the lethal doses death occurred within an hour. Post mortems on the rats that were given an oral dose of 300 mg/kg revealed damage to the liver, spleen and kidneys (no further details). The most conspicuous feature was a black-coloured spleen (Commercial Solvents Corporation, 1960).

In a safety data sheet, an oral LD_{50} for rats of 500 to 1000 mg/kg was given (Allied Corporation, 1983 a).

Other studies gave an intraperitoneal LD_{50} for the mouse of 142 (121 to 166) mg/kg and an oral LD_{50} for the rat of 642 (568 to 725) mg/kg (7-day observation period). The substance was administered as a 2 or 4% aqueous solution. Symptoms observed included dyspnoea, trembling, convulsions and lying on the abdomen. Autopsies of the rats revealed discolouration (blue-violet to blue-black) and enlargement of the spleens (BASF, 1969).

Oral administration of 250 mg/kg of the sulphate in aqueous solution to 5 rats led to slight, temporary cyanosis and inactivity, while 500 mg/kg led to marked cyanosis. Watering of the eyes and convulsions were seen in 2/5 rats, which died within 1 hour of treatment (methaemoglobin positive). The surviving animals were without symptoms 24 hours after administration (ICI, 1984).

In order to study methaemoglobin-formation due to hydroxylamine sulphate, 3 male and 3 female cats were given a single oral dose of 200 mg/kg (as a 2% aqueous solution) by stomach tube. The cats vomited repeatedly, and 2 of them died after 2 days. Autopsy examination of these 2 animals revealed uncharacteristic findings. The methaemoglobin level 4 hours after administration was between 10 and 42%. 2 cats survived a single dose of 50 mg/kg, but salivation occurred, and in one case, cyanosis and vomiting. The methaemoglobin levels were 12 to 22% (BASF, 1981).

The acute toxicity of hydroxylamine sulphate was studied in rats after both subcutaneous and dermal application. In the subcutaneous studies 10 mg/kg was injected, while for the dermal treatment doses of 10 and 500 mg/kg (50% in distilled water) were applied to the skin under a plastic cover. After dermal application, moderate skin irritation was seen (which persisted for 3 to 4 days), as well as, to a lesser extent, scurf and necrosis. Cyanosis was not observed. In all of the treated rats, the methaemoglobin level was slightly but statistically significantly increased 48 hours after exposure, with the highest levels being 3.3% after subcutaneous injection of 10 mg/kg and 4.0% after dermal application of 500 mg/kg. In both of these groups, there was a statistically significant reduction in erythrocytes 14 days after treatment. The number of reticulocytes was only statistically significantly increased after dermal application of 500 mg/kg (130 000/µl; control 70 000/µl). In the surviving animals and those killed after 14 days, the spleens were enlarged and dark in colour, independent of the dose level and route of application (Derelanko et al., 1987).

Groups of 5 male and 5 female New Zealand rabbits received a single dermal application of 500, 1000, 1500 or 2500 mg/kg (concentration not specified) under a gauze cover (24 hours) and were then observed for 14 days. At the highest dose level (2000 mg/kg) 7 rabbits died, from which a dermal LD_{50} for rabbits of between 1500 and 2000 mg/kg was estimated. In all dose groups a severe irritant effect was seen on the skin, together with necrotic changes, cyanosis and hypothermia, the latter effects indicating absorption through the skin. At the low dose levels the irritation cleared up within 3 days of treatment. The rabbits at the high dose levels (1500 and 2000 mg/kg) showed reduced weight gain. After 14 days, the erythrocyte and leukocyte counts were dose-dependently reduced. A reduction in the number of erythrocytes was also seen in the males treated with the lowest dose. In addition, at the two upper dose levels an increase in reticulocytes was observed. The main findings at autopsy were dark-coloured and enlarged spleens, and bloody and necrotic regions on the skin. Histologically, the skin damage was characterized by bleeding, oedema, blister formation, vascular obstruction and massive infiltration (Allied Corporation, 1983 b).

In order to establish whether even lower doses of the sulphate salt also produced toxic effects in rabbits after dermal application, a further study was carried out in which pairs of rabbits were treated with 10, 50 or 100 mg/kg on the clipped dorsal skin. A plastic cover was used at the application site instead of gauze to inhibit loss of the small amount applied. The duration of exposure was again 24 hours. Analysis of the blood 4 days after treatment revealed dose-dependent haemolytic anaemia. A no effect level could therefore not be established at these dose levels either (Allied Corporation, 1983 b).

In a further study on dermal toxicity, groups of 5 male and 5 female rabbits (White Viennese) were treated once with 400 mg/kg as a 50% aqueous preparation applied occlusively to the clipped skin of the back and flank. All of the animals survived the 24-hour exposure period without signs of systemic toxicity or local effects (8-day observation period). Nothing of note was found at autopsy (BASF, 1980).

In another study, 10 female New Zealand rabbits were given 10 mg/kg as a single subcutaneous injection. Other groups of 10 females received 24-hour dermal applications of 1, 10, 100 or 500 mg/kg under a plastic cover or 100, 500 or 1000 mg/kg under a gauze cover. While no deaths were observed at doses of up to 1000 mg/kg under the gauze cover, when the plastic cover was used 7/10 rabbits died at 500 mg/kg and 2/10 at 100 mg/kg. After 24 hours, skin irritation and necroses were seen at the application site and these effects were more severe with the plastic cover. Cyanosis was observed in the rabbits that were treated with 500 or 100 mg/kg under a plastic cover, and to a lesser extent, after subcutaneous injection of 10 mg/kg. While 500 mg/kg under a plastic cover produced a methaemoglobin level of 61% 48 hours after application, only 6.2% methaemoglobin was found at the same time point with the gauze cover, even at a dose of 1000 mg/kg. The number of erythrocytes declined dose-dependently after dermal application under a plastic cover (after 4 days 1.2×10^6 to 4.4×10^6/µl) and after subcutaneous injection (ca. 3×10^6/µl), while after dermal application under a gauze cover it declined only slightly at 1000 mg/kg (4.4×10^6/µl, control 6×10^6/µl). The reticulocyte count increased correspondingly. An increased incidence of Heinz bodies occurred in the erythrocytes of the rabbits 4 days after exposure when the plastic cover was used and after subcutaneous injection. After 14 days the observed effects showed a tendency towards reversibility. Post mortems revealed enlargement and dark discolouration of the spleen in all of the exposed rabbits (Derelanko et al., 1987).

In summary it can be said that dermal application of hydroxylamine sulphate to rabbits causes anaemia even at doses of 10 mg/kg, and can lead to death at doses of 100 mg/kg. From the results of these studies it is evident that the extent of the toxic reactions after dermal application is very much dependent on the method of exposure. The mortality rate and haematological effects were significantly higher after application under a plastic cover than under a gauze cover. In general, the effects on rabbits were more severe than those on rats.

In an inhalation hazard test (20 °C), inhalation of the volatile portion of hydroxylamine sulphate for 8 hours was tolerated by 12 Sprague-Dawley rats without signs of toxicity or deaths (BASF, 1969).

Likewise, no deaths or toxic effects occurred in rats on inhalation of the mist from a saturated aqueous solution of hydroxylamine sulphate for 1 hour (no further details; OTS, 1984).

In a range-finding study prior to a subchronic study, groups of 5 male and 5 female Wistar rats (average starting weights 164.8 and 135.2 g, respectively) were given 25, 100, 400 or 1600 ppm hydroxylamine sulphate (purity ≥99%) in their drinking water for 4 weeks. The actual analytical concentrations however, were clearly lower for the lower doses at 0 to 3 ppm, 18 to 37 ppm and 235 to 325 ppm, respectively. For the top dose the analytical concentration was 1530 to 1760 ppm, approximately the desired value (1600 ppm). In both males and females at the highest dose level, severe haemolytic anaemia and methaemoglobinaemia occurred, with a compensatory increase in erythropoiesis and with related clinical (cyanosis) and pathological (organ weight changes, splenomegaly, haemosiderosis, extramedullary haematopoiesis) findings. These effects were also seen in a less pronounced form in rats in the 400 ppm group (actual concentration 235 to 325 ppm), but only sporadically in the rats in the 100 ppm group (actual concentration ca. 18 to 37 ppm). No treatment-related effects were found in the animals in the 25 ppm group (actual concentration 0 to 3 ppm). Taking the analytical concentrations into consideration, the no effect level in this study was between 18 and 3 ppm for male rats and between 235 and 18 ppm for female rats (BASF, 1989 a).

7.3 Skin and mucous membrane effects

Hydroxylamine

The application of a few drops of a 10% hydroxylamine solution into the conjunctival sac of the rabbit eye caused moderate redness and "trickling tears" (Binz, 1888).

Hydroxylamine hydrochloride

An 80% aqueous preparation of hydroxylamine hydrochloride was applied to the clipped dorsal skin of 2 rabbits (White Viennese) for 1, 5 or 15 minutes (semi-occlusive) or 20 hours (occlusive). No irritation was seen after exposure for 1 or 5 minutes, but after 15 minutes a slight mottled redness occurred which was reversible after 1 week. After a 20-hour exposure, similar findings were recorded which were again reversible after 1 week (BASF, 1975).

Marked oedema and severe clouding of the cornea with a greasy surface layer were evident 1 hour after the instillation of 50 mg into the conjunctival sac of the rabbit eye (2 animals). After 24 hours, oedema, marked clouding, ciliary injection and discharge were seen, and after 8 days marked oedema, marked clouding, iritis, ingrowing vessels and a staphyloma. For the entire period, a grey white discolouration of the conjunctiva and the nictitating membrane was present (BASF, 1975).

When an 8% solution of hydroxylamine hydrochloride was continually dropped into the conjunctival sac of the rabbit eye for 5 minutes, no damage to the cornea occurred. However after the same solution was dropped in for 30 minutes, extreme swelling of the conjunctiva was observed with mydriasis. The cornea was not clouded, but the next day it was cyanotic and severely oedematous (Grant, 1974).

Hydroxylamine sulphate

An 80% aqueous preparation of hydroxylamine sulphate was applied to the clipped dorsal skin of 2 rabbits (White Viennese). The exposure period was 1, 5 or 15 minutes (semi-occlusive) or 20 hours (occlusive). After the 1- and 5-minute applications there were no signs of irritancy, but after 15 minutes slight redness was seen and after the 20-hour exposure severe redness was evident. After 8 days these effects were reversible (BASF, 1969).

Dermal application of a 20% hydroxylamine sulphate solution to the inside of the rabbit ear twice daily on 3 consecutive days did not lead to skin irritation (ICI, 1984).

The instillation of 50 mg hydroxylamine sulphate into the conjunctival sac of the rabbit eye (2 animals) caused slight redness and slight oedema after 1 hour and slight redness, severe oedema and marked clouding of the cornea after 24 hours. These findings had cleared up after 8 days with the possible exception of clouding of the cornea. Accordingly, hydroxylamine sulphate is clearly less irritating to the eye than the hydrochloride (BASF, 1969).

7.4 Sensitization

Hydroxylamine

No information available.

Hydroxylamine hydrochloride

The possible sensitizing effect of hydroxylamine hydrochloride on guinea-pigs after subcutaneous pre-treatment and intracardial challenge has been studied by Soviet researchers. 6-Guinea-pigs were each given 7 mg/kg (corresponding to about 1/10 of the absolute lethal dose) subcutaneously 3 times at intervals of 12 days. A second group of 6 guinea-pigs was given subcutaneous injections of 13.5 mg/kg (equivalent to 1/4 of the absolute lethal dose) 5 times at intervals of 6 days. The treatment was tolerated without signs of toxicity. 12 days after the last subcutaneous injection the animals in the first group were given an intracardial injection of 3.5 mg hydroxylamine hydrochloride (1/20 of the absolute lethal dose). Similarly, those in the second group were injected with 3.5 mg/kg intracardially 6 days after the last subcutaneous injection. 4 of the 12 guinea-pigs died 15 to 20 minutes after the intracardial injection. The remaining animals were restless, shook their heads, scratched themselves, were breathless and suffered convulsions. 4 Guinea-pigs had urinary and faecal discharges 1 to 2 minutes after the injection. The same intracardial dose of 3.5 mg hydroxylamine hydrochloride/kg was tolerated without signs of toxicity by 2 control animals. The authors concluded from this that hydroxylamine hydrochloride had a sensitizing effect (Arnoldova and Speransky, 1963).

Hydroxylamine sulphate

In a dermal sensitization study, 10 guinea-pigs were treated on 12 working days with about 0.1 ml of a 40% aqueous hydroxylamine sulphate solution, painted onto the skin of the shorn left flank in the form of a cross. Distinct skin irritation occurred in all animals, but this cleared up again after 8 to 10 days. 12 days after the last application, the skin of the previously untreated right flank was coated once with ca. 0.1 ml of a 4% hydroxylamine sulphate solution (not a primary irritant) and the skin reaction was noted after 8, 12 and 24 hours. 8 Of the 10 guinea-pigs developed local reactions in the form of redness, oedema and small lumps. Thus hydroxylamine sulphate induced skin sensitization in this study (BASF, 1956).

Hydroxylamine sulphate also had a sensitizing effect in the mouse ear-swelling test. Positive reactions were seen in 33% of the animals (1-chloro-2,4-dinitrobenzene 80%; Gad et al., 1986).

7.5 Subchronic and chronic toxicity

Hydroxylamine

No information available.

Hydroxylamine hydrochloride

When 172 rabbits were injected intravenously with 2.5 mg hydroxylamine hydrochloride/kg, 6 times a week for up to 19 months, they developed methaemoglobin-formation, anaemia and reticulocytosis (Jacobsen et al., 1939).

There were no treatment-related effects on growth or general well-being in rats that were given 333 to 380 mg/kg body weight of the hydrochloride in their milk for 178 days. However, enlargement of the spleen by 4 to 5 times and atrophy of the thyroid gland by about 50% were observed (Riemann, 1950).

In 4 dairy cows that were given 40–60 mg/kg body weight daily in their feed from the second month of pregnancy until they calved or aborted, methaemoglobin levels were 10 to 15% and reductions were seen in the erythrocyte count (on average from 6.4×10^6 to 3.4×10^6) and the haematocrit value (on average from 54.5 to 31.8%; Winter and Hokanson, 1964).

Hydroxylamine sulphate

Four-week-old male Swiss-Webster mice were given 10 or 20 mmol/l (corresponding to ca. 1640 or 3280 mg/l) in their drinking water. One group was treated with the substance for 12 weeks, and two further groups were similarly treated with the substance for 12 weeks and then transferred to pure drinking water for 8 and 18 weeks, respectively. After treatment for 12 weeks, the mice at both dose levels showed a similar degree of spleen enlargement, a reduction in erythrocytes and an increase in leukocytes. These effects were reversible, however, as the parameters were within the normal range in the mice that were subsequently transferred to pure drinking water for 8 or 18 weeks (Yamamoto et al., 1967).

In a subchronic drinking water study based on OECD guideline no. 408, groups of 10 male and 10 female Wistar rats (average initial weights 177 and 136 g, respectively) were given 10, 50 or 250 ppm hydroxylamine sulphate (purity ≥99%) for 3 months in their drinking water ("Milli-Q-Reinstwasser“, completely desalinated, double-distilled water, filtered through fine filters and activated charcoal). Analytical checks at the beginning of the study, after 6 weeks and after 12 weeks gave 9.9, 9.3 and 10.7 ppm for the lower concentration (10 ppm), 48, 46 and 53 ppm for the middle concentration (50 ppm) and 233, 223 and 274 ppm for the upper concentration (250 ppm). The stability of the test substance in "Milli-Q-Reinstwasser“ over 4 days was ascertained before the start of the study. The doses administered correspond to an average daily intake of ca. 0.9, 4 and 21 mg hydroxylamine sulphate/kg body weight. In the 250 ppm group, dark colouration of the urine and a lowering of the erythrocyte and haemoglobin values were seen in both sexes. The females had reduced haematocrit values and the males a reduction in the mean corpuscular haemoglobin concentration. In both sexes there were increases in the mean corpuscular volume, the mean corpuscular haemoglobin value, reticulocytes, Heinz bodies, Jolly's bodies and bilirubin concentration in the serum. Moreover, polychromasia was intensified in both sexes. Autopsy revealed an increase in the absolute and relative spleen weights in both sexes, and increases in the relative liver weight and the absolute and relative adrenal weights in the males. Microscopically, both sexes showed increased haemosiderin storage and sinus dilation with congestion in the spleens, as well as haemosiderin storage in the liver. In both sexes, 50 ppm caused a slight increase in reticulocytes, while in the females this level caused reduced haemoglobin and erythrocyte leves, a marginal increase in the mean corpuscular volume and a negligible increase in polychromasia. Histologically, moderately increased haemosiderosis was found in the spleen in both sexes. No treatment-related effects were found at 10 ppm. The no effect level for both sexes was between 10 ppm (equivalent to ca. 1 mg/kg body weight/day) and 50 ppm (equivalent to ca. 4 mg/kg body weight/day) in this study (BASF, 1989 b).

7.6 Genotoxicity

7.6.1 In vitro

Hydroxylamine

In a study on the effects of hydroxylamine on human chromosomes in vitro (fibroblast cultures from subcutis biopsies) it was found that 0.3 mmol/ml (equivalent to 10 µg/ml) caused chromosome breaks after an exposure period of 36 hours, the break frequency being increased mainly in the centromere region (Engel et al., 1967).

Studies were conducted to investigate the mutagenic effect of hydroxylamine on chromosomes from cells of various organs (kidney, lung and embryonal fibroblasts) from hamsters of the subspecies Phodopus sungorus. Negative results were obtained at concentrations of 0.01, 0.025 and 0.05 mg/ml for 24 or 48 hours. Higher concentrations could not be tested because 0.05 mg/ml was already cytotoxic (Sokova et al., 1970).

Hydroxylamine was studied for cytotoxic effects and for its ability to induce forward mutations and chromosomal aberrations in cultured cells of the CHO clone and the CHO-K1 subclone from the Chinese hamster. Test concentrations of 40 to 150 µg/ml were used, with a 16-hour incubation period. The hydroxylamine concentration at which a 20% cell survival rate was reported was 150 µg/ml. While no mutants were induced at this concentration, a dose-dependent increase in chromosome breaks and exchanges was found in the 80 to 120 µg/ml concentration range. On the basis of these results, hydroxylamine was classified as one of a group of compounds (including caffeine) that are effective in inducing chromosomal aberrations but are only slightly active in inducing forward mutations (Kao and Puck, 1969).

In the characterization of mutants by genetic tests it was found that the ad-3-mutations in *Neurospora crassa* induced by treatment with 1 M hydroxylamine (equivalent to 33 mg/ml) for 4 to 12 hours were point mutations that were mainly due to base transitions from GC to AT (Malling and de Serres, 1971).

Other authors have observed direct effects of hydroxylamine on isolated mammalian and phage DNA, with a clear reduction in the viscosity and the sedimentation coefficient as well as double-stranded fragments with molecular weights of approximately 5×10^5. Lower hydroxylamine concentrations led to considerable denaturation, which the authors thought was probably due to an attack on cytosine (Bendich et al., 1963).

In a more recent study, which is only available in the form of a short report, hydroxylamine caused no thioguanine-resistant mutants in Chinese hamster V79 cells (Taylor et al., 1985).

In a study of the mutagenic effect on transfected DNA from phage phi 105 of Bacillus subtilis, hydroxylamine was less effective than methyl methane sulphonate, ethyl methane sulphonate and N-methyl-N-nitro-N-nitrosoguanidine by 1 to 2 orders of magnitude (Mahna, 1984).

Hydroxylamine hydrochloride

The effect of treatment with 5 and 25 µg hydroxylamine hydrochloride/ml for 24 and 48 hours on the chromosomes of Chinese hamster cells cultured in vitro (cell line 11dFAF28) was studied. The treatment led to chromosome and chromatid breaks, which were often found in the guanine-cytosine-rich regions of the centromere on chromosome pairs 1, 2 and X (Somers and Hsu, 1962).

Hydroxylamine hydrochloride at a concentration of 7.2×10^{-5} M also caused structural chromosome changes (13%, control 5%) in another Chinese hamster cell line (derived from a methyl-cholanthrene-induced tumour, no further details) after incubation for 72 hours. The mitotic index was 2.8% (control 4.8%). Concentrations of more than 10^{-4} M were toxic (Borenfreund et al., 1964).

In a study of the localization of chromosome breaks in the Muntjak (an Indian wild deer) after treatment of lymphocyte cultures with 50 µg/ml for 18 hours, it was established that hydroxylamine hydrochloride acted upon the GC-rich centromere and the secondary constrictions of the chromosomes. Here, constrictions, gaps and chromatid breaks occurred. The frequency of breaks in the centromere region of chromosome pairs 1, 2 and X was significantly increased (Gupta and Sharma, 1981).

In a further study, the same authors looked at the effect of the hydrochloride on cell-cycle kinetics and on the frequency of sister chromatid exchange (SCE) in Muntjak lymphocytes. The chemical was added at concentrations of 25 or 50 µg/ml (in each case over a period of 1 hour) at various times after the start of the culture, i.e. after 1, 30, 44 and 46 hours. The cultures were evaluated after 48 hours. The cells treated at various times showed a reduction in metaphases, and there was a slight increase in SCE frequency in the cultures treated after 1 hour, whereas at later treatment times no differences from the control cultures were found. According to the authors these findings indicate that the affected cells cannot pass through more than 2 cycles (Gupta and Sharma, 1982).

In studies on the induction of SCEs in the Chinese hamster V79 cell line, hydroxylamine hydrochloride caused a slight but dose-dependent increase in SCEs after treatment for 1 hour at 10^{-4} M (equivalent to 7 µg/ml) and above. The concentration range tested was from 10^{-5} to 5×10^{-3} (equivalent to 0.7 to 350 µg/ml; Speit et al., 1980).

When the cell line R-1CA (derived from a primary culture of monkey kidney cells) was treated with 6.25, 12.5, 25, 50 or 100 µg hydroxylamine hydrochloride/ml for 8, 24 and 48 hours, changes in the nucleolus (hypertrophy) were evident at the 3 higher concentrations 48 hours after the beginning of treatment. After 8 hours acentric chromosome fragments and micronuclei were already visible during mitosis at these 3 concentrations (Mironescu et al., 1969).

In human leukocyte cultures, hydroxylamine hydrochloride caused no significant chromosome damage at concentrations of 67 to 500 µmol after a 72-hour incubation. However, this may have been due to technical errors, as the leukocytes were contaminated with erythrocytes, and 2 to 3% contamination with haemoglobin would be sufficient to decompose the amount of hydroxylamine used (Oppenheim and Fishbein, 1965).

In another study in which there was evidently no contamination with haemoglobin, chromosome damage occurred dose-dependently in human leukocytes after incubation for 2 to 12 hours (5 and 25 μg/ml). The damage was characterized mainly by constrictions and gaps, while breaks and exchanges were seen less frequently. Concentrations of 50 and 100 μg/ml were toxic after longer incubation and led to inhibition of mitosis. Hydroxylamine hydrochloride did not induce sister chromatid exchange in this test nor in a test with HeLaS3 cells (Brøgger, 1971).

Hydroxylamine hydrochloride concentrations of 1 and 1.5 M (equivalent to 70 and 105 mg/ml or 7 and 10.5%, respectively) caused an increase in the frequency of mutations in the yeast strains TD114, S2615B and S2614C after 2 hours of treatment, and an increase in the frequency of recombinations in the strains SC13-22SBTD and TDSC22. The ability of hydroxylamine hydrochloride to induce recombinations was interpreted by the authors as an indication that the changes it induced in DNA could be repaired (Putrament and Baranowska, 1971).

Hydroxylamine hydrochloride was tested for recombinogenic activity in Saccharomyces cerevisiae D3. In this study, a 0.02% solution in DMSO was positive with and without metabolic activation (S9-mix from Aroclor 1254-induced rat liver; Simmon, 1979 a).

Hydroxylamine hydrochloride caused an increase in novobiocin resistant mutants in Haemophilus influencae at concentrations ranging from 0.14 to 1.25 M (equivalent to 0.7 to 87 mg/ml) after exposure for between 5 and 30 minutes (Kimball and Hirsch, 1975).

No mutagenic activity was detected when hydroxylamine hydrochloride (5000 μg/plate) was tested in the Salmonella/microsome test in strains TA 1535, TA 1537, TA 98 and TA 100 after metabolic activation (McCann et al., 1975).

In a further Salmonella/microsome test in strain TA 1535 hydroxylamine hydrochloride likewise proved not to be point-mutagenic at concentrations of 25 to 250 μg/plate, either with or without metabolic activation. However, positive activity was seen at a dose of 500 μg/plate in a DNA-polymerase-deficient strain of *Escherichia coli* (Rosenkranz and Poirier, 1979).

Again, hydroxylamine hydrochloride did not induce point mutations in a Salmonella/microsome test in the strains TA 1535, TA 1536, TA 1537, TA 1538, TA 98 and TA 100 at a dose of 1000 μg/plate, either with or without metabolic activation (S9-mix from Aroclor 1254-induced rat liver; Simmon, 1979 b).

A study of hydroxylamine hydrochloride (as part of the National Toxicology Program) in an Ames test in the *Salmonella typhimurium* strains TA 1535, TA 1537, TA 1538, TA 98 and TA 100 with and without metabolic activation similarly gave negative results. Negative results were also obtained in *Escherichia coli* WP2uvrA, in 2 mouse lymphoma tests and in a test for the induction of DNA repair (UDS; NIH/NTP, 1984).

In another study, for which only an English abstract is available, hydroxylamine (anion not specified) was positive in the *Salmonella typhimurium* strain TA 104 at a concentration of 10^{-4} mg/ml. The mutagenic effect decreased with increasing concentration, because of increasing toxicity (no further details; Menevse et al., 1984).

Hydroxylamine hydrochloride was tested in the DNA-repair test in primary hepatocyte cells originating from male Fischer-344 rats. The cells were incubated with the chemical at concentrations of 1 and 2 mg/ml for 18 to 20 hours at 37 °C. While 2 mg/ml was cytotoxic, 1 mg/ml was without effect (Williams et al., 1982).

In T4-phages, various concentrations of hydroxylamine hydrochloride had variable effects. At high concentrations (ca. 1 M) mutation of the phages occurred with little inactivation, while at low concentrations (ca. 10^{-2} M) the phages were inactivated without mutants being observed (Freese et al., 1961). Later it was ascertained that this mutagenic effect depended on a specific reaction of hydroxylamine hydrochloride with cytosine (Freese, 1963).

In order to test whether the inactivating effect on the DNA could be confirmed at lower concentrations, further studies were carried out with the transformation system in *Bacillus subtilis*, which enabled the determination of both inactivation and mutation in the same DNA. The authors found that hydroxylamine hydrochloride actually had two different effects on the DNA, with one mainly mutagenic effect at high concentrations (ca. 1 M) and one mainly inactivating effect at low concentrations (10^{-2} to 10^{-6} M). During treatment with the hydrochloride, the mutation rate increased linearly with time (2 to 6 hours) and in relation to the concentration (2, 1, 0.5 and 0.1 M). Likewise, the inactivation was both concentration-dependent (1 to 10^{-6} M) and time-dependent (within 1 to 3 hours; Bautz-Freese and Freese, 1964).

Later it was reported that the inactivating effect was indirectly caused by metabolic products [Nitroxyl (HNO), hydrogen peroxide (H_2O_2) and various radicals] from the reaction of hydroxylamine and O_2 (Freese and Bautz-Freese, 1965).

When a dose of 2×10^{-2} M hydroxylamine hydrochloride (equivalent to 1.4 mg/ml) was applied to culture plates containing bacteriophage S13 with *Escherichia coli* as host cells (incubated at 37 °C for 6 hours), all four bases (guanine, adenine, cytosine and thymine) were mutated. In contrast, when the phage was treated alone without the host cells, only cytosine was affected (Tessman et al., 1965).

Hydroxylamine hydrochloride has also been tested in the L5178Y mouse lymphoma test with and without metabolic activation (S9-mix from Aroclor 1254-induced rat liver). In three test series there was a dose-dependent positive effect ($p \leq 0.05$) in the concentration range 31.3 to 160 µg/ml without activation. With metabolic activation, the effect was less uniform. In three experiments, positive results were found in the dose range 220 to 300 µg/ml ($p \leq 0.05$) and in a fourth experiment, the result was negative in the range 150 to 300 µg/ml. The weaker mutagenic effect of hydroxylamine hydrochloride after metabolic activation was attributed to its higher toxicity (Myhr and Caspary, 1988).

Various chemicals were studied in the cell transformation test with primary embryo cells from the Syrian golden hamster. The incubation period ranged from a few hours up to 8 days, and the test was conducted both with and without metabolic activation (hamster liver homogenates or hamster hepatocytes). The highest hydroxylamine hydrochloride concentration tested (100 µg/ml) had no cell-transforming activity in this test system, either with or without metabolic activation (Pienta, 1980).

Hydroxylamine hydrochloride was also negative in two further cell lines in the in vitro cell transformation test. Concentrations of 0.2, 1.0, 5.0 and 25.0 µg/ml caused no transformations in BALB/3T3 mouse cells after 3 days of treatment. Similarly, concentrations of 0.01, 0.1, 1.0, 10.0 and 100.0 µg/ml had no cell-transforming effect on embryonal Chinese hamster cells after 8 days of treatment (Dunkel et al., 1981).

While hydroxylamine hydrochloride was described as having cell-transforming activity in a test involving rat embryo cells

($2FR_4 50$) infected with Rauscher Leukaemia virus after treatment for 72 hours at 24 µg/5.2×10^4 cells (Traul et al., 1981), no cell-transforming activity was found in a series of National Toxicology Program studies in which the same cells were used (no further details; NIH/NTP, 1984).

Hydroxylamine sulphate

Hydroxylamine sulphate was studied in the cell transformation test with BALB-3T3-clone-A-31 mouse embryo cells, in a 24-hour exposure at 36 °C with and without metabolic activation. In the absence of metabolic activation, concentrations of 30 and 3 µg/ml caused a significant increase in the transformation frequency of type 3 foci ($p<0.01$), whereas 10 µg/ml did not ($p=0.05$). The authors concluded from this that hydroxylamine sulphate has cell-transforming properties (Microbiological Associates, 1981 a).

In further studies of the same design, concentrations of 100, 30 and 10 µg/ml were tested with the addition of a metabolic activation system. The transformation frequency of type 3 foci was only significantly increased at 30 µg/ml ($p=0.02$) and not at the higher (100 µg/ml) or lower (10 µg/ml) concentrations ($p>0.05$; Microbiological Associates, 1981 b).

7.6.2 In vivo

Hydroxylamine

In *Drosophila melanogaster*, the induction of certain ‘miniature‘ mutations by hydroxylamine at a concentration of 0.03 M in the feed (equivalent to ca. 1000 ppm) was studied. It was found that after treatment at the late larval spermatogonia or pre-meiotic spermatocyte stages, mutations of the appropriate gene were induced (Shukla, 1972).

In fruit flies treated with hydroxylamine from day 3 after egg laying, at concentrations of 0.01 to 0.04 M (equivalent to 330 to 1320 ppm in the feed), locus-specific mutated phenotypes were microscopically observed (Jain and Shukla, 1972).

In a study of the induction of lethal mutations in *Drosophila melanogaster*, delayed mutagenesis occurred in the F_2 or F_3 generations after treatment with 0.1 M hydroxylamine in the feed (equivalent to 3300 ppm) for 48 hours. It was shown that the incidence of lethal mutations in the F_2 generation was slightly but not significantly higher than in the controls, while in the F_3 generation the increase was significant (Shukla and Auerbach, 1979).

In a pilot study of a new Drosophila spot test, 72-hour-old *Drosophila melanogaster* larvae were treated with 0.045 M hydroxylamine (ca. 1500 ppm in the feed) for 48 hours and examined for the induction of gene mutations and somatic recombinations. No clear mutagenic or somatic recombination-inducing effect was reported (Graf et al., 1983).

Hydroxylamine hydrochloride

The injection of 0.25 µl of a 100 mmol solution of hydroxylamine hydrochloride (corresponding to 1.75 µg) into the haemocoele of male fruit flies (*Drosophila melanogaster*) had a cytotoxic effect on spermiogenesis. The hatching rate was reduced by 50% compared with the controls, but cytological examination of the damaged eggs revealed no dominant lethal mutations. After injection of 125 mmol (equivalent to 2.19 µg), the mutation rate was approximately doubled in the X- and 2nd chromosomes of the offspring, but this was not statistically significant. No chromosome aberrations, chromosome breaks, translocations, "minutes" or deletions occurred. There were no sterile Y-chromosomes or hyperploid females. No X-recessive or mosaic X-recessive mutations were observed in the F_1 and F_2 generations (Fahmy and Fahmy, 1970).

After 48 hours feeding of 45 mM hydroxylamine (anion not specified) to 72-hour-old Drosophila larvae, there was a significant increase in single spots but not in twin spots (short report, no further details; Würgler et al., 1983).

A single abdominal injection into male locusts of 0.05 ml hydroxylamine hydrochloride as a 0.1 M solution (equivalent to 350 µg) caused chromosome aberrations (breaks, fragmentation, bridge formation) in the spermatocytes 36 hours later, effects which were not observed in the controls (Bhattacharya et al., 1986).

Hydroxylamine sulphate

In studies on the mutagenicity of hydroxylamine sulphate in the micronucleus test, 4 male and 4 female CD-1 mice were given 15.6 or 125 mg/kg by stomach tube, administered as two doses separated by an interval of 24 hours. Evaluation of polychromatic erythrocytes from mice killed 6 hours after the second treatment (30 hours after the first dose) showed that no increase in micronuclei was triggered by either dose in either the males or the females (Brusick and Stetka, 1984; Litton Bionetics Inc., 1979/1980).

The effects of hydroxylamine sulphate were studied in a modified dominant lethal test (Epstein and Röhrborn, 1971) in male and female ICR/Ha Swiss mice. The males were given a single intraperitoneal injection of 102 mg/kg (7 mice) or 112 mg/kg (9 mice), and were then each paired with 3 untreated females for 1 week. The females were exchanged for new untreated females weekly for a period of 8 weeks. The parameters studied (early foetal deaths, pre-implantation losses) were within the range of the control levels (review article, no further details; Epstein et al., 1972).

In contrast, hydroxylamine sulphate induced chromosome breaks in *Drosophila melanogaster* when administered to the larvae at a concentration of 0.5% in their feed. There were effects on 21.43% of the treated larvae compared with 0% of the controls (Parkash and Miglani, 1978).

7.7 Carcinogenicity

Hydroxylamine

No information available.

Hydroxylamine hydrochloride

In a long-term study, a total of 78 male AKR mice, 60 female C3H mice and 85 male Swiss mice were given pelleted feed to which 1 or 2% by weight of hydroxylamine hydrochloride had been added. Treatment continued for their entire lifetime. While tumours developed in most C3H controls (no further details), there was a considerable reduction in the spontaneous tumour incidence in the group treated with 2% of the hydrochloride (0% after 15 months compared with 20% in the controls). In the AKR mice the treatment led to a dose-dependent increase in survival time. A positive influence on the survival time was also established in the C3H mice given 1% by weight. In the male Swiss mice no relevant effects were found in either dose group. No substance-induced tumours were observed in any of the mouse strains (Harman, 1961).

Hydroxylamine sulphate

5 male Swiss-Webster mice and 10 female C3H/HeN mice were given hydroxylamine sulphate in their drinking water for 52 weeks at concentrations of 10 and 20 mmol/l (equivalent to 1640 and 3280 mg/l, respectively). At the end of the treatment period the spleens were about 10-fold heavier than in the controls and nearly

50% of the mice had areas of bone formation in the spleen. No treatment-related tumours were observed. In the females no mammary tumours were found after 1 or even 2 years compared with a 100% incidence in 10 untreated females (Yamamoto et al., 1967).

The effect of hydroxylamine sulphate on the mammary glands, ovaries and pituitary was studied in female C3H/HeN mice after administration of the chemical in the drinking water as a 10 mM solution (equivalent to 1640 mg/l) for 50 to 70 weeks. Treatment was begun at different ages (6, 13, 18, 24 and 28 weeks) in various experiments. After treatment began at the age of 6 weeks (98 mice) there was a clear delay in development of the mammary glands and ovaries. The frequency of spontaneous mammary tumours was significantly reduced (31%) compared with controls (100%). When treatment began at 13 weeks (30 mice), the mammary tumour rate was 40% compared with controls (100%). Up to the age of 32 weeks a clear effect on the ovaries was shown in cross section (5 mice), in the form of a reduction in Corpora lutea and an increase in the number of mature follicles. After treatment at the age of 18 or 24 weeks (30 mice in each case), there was a tendency towards increased tumour frequency. In the mice without tumours, the prolactin content in the pituitaries was significantly increased in the treated mice compared with the controls, whereas no such difference was observed in the mice with tumours. These results suggest that hormonal factors played a role in suppressing the development of mammary tumours (Evarts and Brown, 1977).

The effect of hydroxylamine sulphate on mammary tumours induced by 7,12-dimethylbenz(a)anthracene (DMBA) was studied in Sprague-Dawley rats, which were given the chemical in the drinking water as a 10 mM solution. When the drinking water administration began 2 weeks after DMBA-treatment (11 mg/rat by stomach tube), the number and weight of mammary tumours per rat was reduced ($p \leq 0.05$) and the average latency period was increased ($p \leq 0.01$). When administration began 2 weeks before DMBA-treatment, these parameters were unchanged, but the normal lobular structure of the mammary glands was clearly protected against the destructive effect of DMBA (Evarts et al., 1979).

A further study assessed the influence of hydroxylamine sulphate on the tumour incidence and the life span of mice. Groups of 40 to 50 male and female mice (age at beginning of study 6 weeks, strain C3H/HeJ(+) or C3H/HeN) were given hydroxylamine sulphate in their drinking water for 105 to 123 weeks. The concentration was 1640 mg/l (10 mM). The reductions in the mammary tumour incidence reported by Yamamoto et al. (1967) and Evarts and Brown (1977) were confirmed in C3H/HeN mice, but not in the C3H/HeJ(+) mice. Instead, other tumours were significantly increased (haemangiosarcomas of the spleen, $p<0.01$, and of the lymph nodes, $p<0.05$, as well as lymphomas, $p<0.05$). The increase in survival time due to hydroxylamine administration, as described by Harman (1961), could not be confirmed (attributed by the authors to better conditions of hygiene; Stenbaeck et al., 1987).

7.8 Reproductive toxicity

Hydroxylamine

In a study of the embryotoxic effects of hydroxylamine in New Zealand rabbits, doses of 50 to 20 μg/application site were injected directly into the embryonal abdominal cavity (coelom) on day 12 of pregnancy in order to avoid methaemoglobinaemia in the dams. Higher doses led to a high degree of foetal resorptions. Cranio-facial and limb anomalies were observed in the foetuses treated with the lower doses. Histological examination revealed mesenchymal necroses in the neural tube, in the limb buds and in the facial region, which developed within 2 hours. From these results the author concluded that hydroxylamine has a potent embryotoxic effect (DeSesso, 1980). However, the significance of these findings is considerably reduced because of the unusual administration route.

According to a study that is only available as an abstract, rabbits were given subcutaneous or intrauterine injections of hydroxylamine (dose not specified) on day 12 of pregnancy. 4 hours later, the embryos were examined under a light microscope. Hydroxylamine caused cell death in the limb buds. Simultaneous injections of propyl gallate (dose likewise not specified) prevented this effect (DeSesso, 1987, 1980). These results are also only of limited value.

Hydroxylamine hydrochloride

Pregnant Wistar rats were injected intraperitoneally with 50 to 2000 mg hydroxylamine hydrochloride/kg on day 9, 10, 11 or 12 of pregnancy, and were killed on day 21. Despite toxic effects in the dams and the foetuses, which were not described in more detail by the authors, no malformations were seen in the foetuses (Murphy and Chaube, 1964).

In a further study, 6- to 8-week-old pregnant Wistar rats were given single intraperitoneal injections of 3 to 15, 29 to 35 or 47 to 59.3 mg of the hydrochloride/kg body weight on day 9, 10, 11 or 12 of pregnancy. The foetuses were delivered on day 21 post-conception, and were examined for malformations. The numbers of resorption sites and dead foetuses were also ascertained. Doses of 47 to 59.3 mg/kg were lethal to 8/12 dams. The LD_{50} value for the pregnant rat estimated on day 11 was given as 59 mg/kg. In the live foetuses from the surviving dams (observed for 21 days) there were no anomalies (not teratogenic). However, administration of 47 mg/kg on day 12 of pregnancy resulted in 32% resorption sites. The embryotoxic effect was attributed to the maternal toxicity (Chaube and Murphy, 1966).

Other authors found that intravenous injection of the hydrochloride (10 mg/kg body weight) on day 3 or day 8 post-conception induced teratogenic effects in crossbred rabbits of variable size and origin (number of animals not specified). Macroscopic examination of the foetuses and determination of body length took place on either day 11 or day 17 post-conception. The Corpora lutea of each female were counted and the percentages of live normal, live malformed, and dead or resorbed embryos were calculated. On day 17, 90% of the foetuses from dams treated on day 3 were either malformed or dead. All of the foetuses of dams treated on day 8 were dead and resorbed. For the dams treated on day 3 and killed on day 11, over 50% of the foetuses showed no anomalies. The proportion was similar in foetuses of dams treated on day 8 and killed on day 11. The observed anomalies occurred mainly in the region of the head. After administration on day 3 and day 8 of the study the foetuses had a reduced body length compared with controls on day 11 post-conception, but were larger than the controls on day 17 post-conception (Zimmermann and Gottschewski, 1966). However, there are considerable shortcomings in the experimental methods (e.g. non-homogenous population of experimental animals, administration only once and intravenously, no clear statement on maternal toxicity) and these findings are therefore of limited significance.

Pregnant New Zealand rabbits were treated with hydroxylamine hydrochloride either intravenously or subcutaneously on day 12 of pregnancy. The doses used were 50 mg/kg body weight intravenously (2 rabbits) and 600, 450, 300 or 200 mg/kg body weight subcutaneously (1 rabbit at each dose). A further 7 rabbits were treated subcutaneously with 50 mg/kg body weight. The rabbits that were given 50 mg/kg body weight intravenously showed clinical signs of methaemoglobinaemia and died within 10 minutes. When 600 to 200 mg/kg body weight were injected subcutaneously, the rabbits died after 5 to 26 minutes, while of those given 50 mg/kg, one died within 4 hours, and a second died during the following night. The remaining rabbits were killed after 5 hours (3 animals) or 8 hours (2 animals), and their uterine contents were examined. At both times the uteri appeared cyanotic. The embryos collected at 5 hours were all alive, but showed dilations of the anterior and posterior cardial veins. In addition, petechial haemorrhages were seen in the cranial region in 20 out of 21 embryos, and red fluid was found in the pericardial cavity in 7 out of 21 embryos. The 16 embryos examined after 8 hours were all dead. All of the above effects were considered to be a consequence of the maternal toxicity of hydroxylamine hydrochloride. To overcome this, a further study was carried out in which the chemical was injected directly into the chorionic cavities of the embryos after laparotomy of the does. The doses used were 200 µg (4 rabbits), 100 to 175 µg (4 rabbits), 75 µg (6 rabbits), 50 µg (6 rabbits) and 25 µg (5 rabbits) per embryo. Alternate embryos, which were injected with saline, served as controls. Doses of 75 µg and above were embryolethal, causing a resorption rate of 91%. After doses of 75 to 200 µg/embryo, the surviving 5 foetuses showed a statistically significant reduction in body weight and, in addition, 2 exhibited malformations of the skull and sternum. After doses of 25 and 50 µg/embryo, the resorption rate was about 60%, but the surviving foetuses showed neither reduced body weights nor structural malformations. Microscopic examination revealed necrosis of the limb buds, an effect which was also observed after the application of the base. Thus hydroxylamine hydrochloride proved to be embryotoxic on intracoelomic injection, even at doses as low as 25 µg/embryo, but was not shown to be teratogenic (DeSesso and Goeringer, 1990). The value of these findings in practice is questionable.

The effect of hydroxylamine hydrochloride on the embryonic development of chicks has also been studied. On day 3 of incubation, a window was bored in the eggs and closed off with transparent adhesive tape. On day 3, 6, 7, 10, 12 or 13 of incubation, 0.1 ml of a hydroxylamine hydrochloride solution was injected into the air space. Three different doses were administered on day 3: 0.25 to 1.0 mg, 0.012 to 0.1 mg and 0.001 to 0.006 mg. On days 6, 7, 10 and 12 the embryos were treated with 0.006 mg and on day 13 with

0.012 mg. On day 14 of incubation, all of the embryos were killed and macroscopically examined for malformations. The incidence of malformations in the controls was 10.6%, compared with 15% in the groups treated with the hydrochloride. From these results the authors concluded that hydroxylamine hydrochloride had a very weak teratogenic effect after administration to chicken embryos on days 6 to 13 of incubation, resulting in an increased frequency of short-wingedness (brachymelia). There were practically no differences between the control and treated groups with regard to other malformations (Stoll et al., 1967 a, b).

Hydroxylamine hydrochloride was administered to 4 pregnant dairy cows (in their grain) at doses of 40 to 60 mg/kg body weight, with a quarter of the dose given 4 times daily. Treatment began after the second month of pregnancy and lasted until its end or until the day of autopsy. The calves were dissected within 24 hours of birth. In the mothers the erythrocyte count and the haematocrit level were reduced and the methaemoglobin level was 10 to 15%, yet the course of pregnancy was unaffected. None of the calves had malformations (Winter and Hokanson, 1964).

Hydroxylamine sulphate

No information available.

7.9 Effects on the immune system

No information available.

7.10 Neurotoxicity

Hydroxylamine hydrochloride

As early as 1889 it was reported that a rabbit given a subcutaneous injection of 55 mg hydroxylamine hydrochloride/kg became restless and developed clonic spasms, strychnine-type convulsions and nystagmus within 10 minutes of treatment. The author concluded that the administration of very high doses to warm-blooded animals led to paralysis of the entire central nervous system immediately after convulsions (Lewin, 1889).

In rats, violent convulsions were seen before death after the oral administration of lethal doses of hydroxylamine hydrochloride (OTS, 1984; Gross and Smith, 1985; BASF, 1975).

7.11 Other effects

Hydroxylamine

It has been reported that the incubation of bovine or porcine liver carboxylesterases (in vitro) with 0.25 M hydroxylamine (equivalent to 8.25 mg/ml) causes rapid inactivation of these enzymes (Runnegar et al., 1968).

In contrast to the above-mentioned inhibitory effect on enzymes, a concentration of 5×10^{-2} M hydroxylamine (equivalent to 16.5 mg/ml) increased the aromatase activity of microsomal cytochrome P450 from human placenta in vitro. The activity was 140% of that of the control (Zachariah et al., 1976).

Another study has shown that a 2 mM concentration of hydroxylamine (equivalent to 66 µg/ml, 10-minute incubation at 37 °C) inactivates guanylate cyclase in the brain cortex of the rat (Deguchi, 1977).

Hydroxylamine hydrochloride

Subcutaneous injection of 0.015 mmol/mouse (equivalent to ca. 50 mg/kg) on 4 consecutive days led to inhibition of catalase activity in the blood and liver, but this had returned to normal after 6 days (Adams and Roe, 1953).

In a study of the effect of hydroxylamine hydrochloride (at a concentration range of 1.6×10^{-5} to 4.6×10^{-3} M, equivalent to 1.1 to 320 µg/ml) on cytochrome oxidase from bovine heart, a reversible inhibitory effect was ascertained (Yoshikawa and Orii, 1970).

The therapeutic usefulness of hydroxylamine hydrochloride in the treatment of cyanide poisoning was studied in mice. After intraperitoneal injection of sodium cyanide the LD_{50} was 6 mg/kg. Intraperitoneal injection of 76 mg hydroxylamine hydrochloride/kg, 10 minutes before the cyanide injection, resulted in an LD_{50} value for the cyanide of 8.8 mg/kg. The protective effect was only about half that of sodium nitrite (Kruszyna et al., 1982).

20 male Wistar rats (248±20 g) were given 0.3 g hydroxylamine hydrochloride/l drinking water (4.3 mM) for periods of from 1 to 63 days. At the end of the study the cumulative amount was 755±121 µmol/kg body weight (equivalent to 53±8 mg/kg body weight). Several biochemical parameters were measured after 1, 49 and 63 days of treatment in whole brain homogenates, glial cell homogenates and muscle homogenates (5 rats each). After 63 days the succinate dehydrogenase activity in the total brain homo-

genate was marginally increased, and the creatine kinase activity in the glial homogenate was clearly increased ($p<0.01$). In the muscle homogenate the acetylcholinesterase activity was only slightly increased after one day ($p<0.05$); the creatine kinase activity was increased after 1 and 63 days of treatment ($p<0.01$), but not after 49 days of treatment (Savolainen, 1984). The significance of these findings is not clear.

Intravenous injection of hydroxylamine hydrochloride caused a dose-dependent drop in blood pressure in anaesthetized dogs and cats. In cats, after injections of 2, 4, 6 and 8 mg/kg the blood pressure dropped to 57, 61, 63 and 53% of the initial value, respectively. The effect was more slowly reversible than that after treatment with 6 mg sodium nitrite/kg body weight (Kruszyna et al., 1984).

Hydroxylamine sulphate

No information available.

8 Experience in humans

Hydroxylamine

It has been reported that at the end of a working day, employees have blood methaemoglobin concentrations of up to 25% (OTS, 1984).

Hydroxylamine hydrochloride

There is a report of acute poisoning due to oral ingestion of hydroxylamine hydrochloride. A woman who had allegedly mistakenly drunk about 2 gulps of a solution of the hydrochloride showed no immediate symptoms but, 2 hours later, developed nausea, vomiting and cyanosis of the nails. There was also a moderate progressive haemolytic anaemia with leukocytosis and an increased rate of blood sedimentation. The convalescence lasted 5 weeks. The thrombocyte count, bleeding time and coagulation time were normal (Makotcenko and Rutstejn, 1974).

Exposure to hydroxylamine hydrochloride reportedly led to eczematous dermatosis in 5 out of 13 employees. In diagnostic skin tests all 5 people reacted to hydroxylamine hydrochloride (Gobbi, 1970).

It has been reported that contact of the skin with hydroxylamine hydrochloride leads to irritation and itching, which intensifies during the night. After longer exposures, allergic dermatitis of the erythematous-squamous type can develop, with the body and the lower third of the forearm predominantly affected. The symptoms generally appear 1 to 2 weeks after work commences, but sometimes they may not appear until much later (up to 3 to 5 years later; ILO, 1983).

Contact dermatitis developed in 8 out of 20 people employed in the production of hydroxylamine hydrochloride. In diagnostic tests, 7 of the 8 people had a positive skin reaction to 0.1 and 1% solutions of the hydrochloride (Folesky et al., 1971).

In a further report, the cases of 11 people from the same factory who had all worked with hydroxylamine hydrochloride and hydroxylamine sulphate were described. 4 had eczematous skin on the hands and affected nails, while 6 complained only of subjective sensations such as tingling and itching. 6 of the 11 reacted positively in tests with 2 and 1% aqueous solutions of hydroxylamine sulphate, and 3 out of 11 also reacted to a 0.1% solution. However, 14 out of 34 control subjects also reacted positively (Pellerat and Chabeau, 1975).

Hydroxylamine sulphate

A case has been reported of a worker who was involved with various colour developers, one of which contained hydroxylamine sulphate as a component. 5 months after the beginning of his employment, he developed chronic hand eczema with partial loosening of the fingernails. Fungal infection was excluded. In patch tests with the colour developers only the one containing hydroxylamine sulphate produced a positive reaction, and of the components, only hydroxylamine sulphate (2 and 5% in water) provoked a positive reaction after 24 hours and a strongly positive reaction after 96 hours (Goh, 1990).

9. Threshold limit values

No information available.

References

Adams, D.H., Roe, F.J.C.
The action of some chemical substances on mouse liver catalase activity in vivo
Br. J. Cancer, 7, 509–518 (1953)

Allied Corporation
TSCA sec. 8(e) submission 8HEQ-0783-0486
The toxicity of hydroxylamine and its salts
Literature review, Office of Toxic Substances, Washington DC, U.S. Environmental Protection Agency (1981)

Allied Corporation
Hydroxylamine sulfate: product safety data sheet (1983 a)

Allied Corporation
TSCA sec. 8(e) submission 8EHQ-0983-0486, Follow-up
Hydroxylamine sulfate toxicity study; acute dermal toxicity study of hydroxylamine sulfate
Office of Toxic Substances, Washington DC, U.S. Environmental Protection Agency (1983 b)

Amarger, N., Alexander, M.
Nitrite formation from hydroxylamine and oximes by *Pseudomonas aeruginosa*
J. Bacteriol., 95, 1651–1657 (1968)

Arnoldova, K.A., Speransky, N.N.
Some aspects of the effect produced by muriatic hydroxylamine on the animal body (english translation)
Gig. Tr. Prof. Zabol., 7, 38–42 (1963)

BASF AG, Gewerbehygienisch-pharmakologisches Institut
Bericht über die orientierende biologische Prüfung von Hydroxylammoniumsulfat
Unpublished report (1956)

BASF AG, Gewerbehygienisch-pharmakologisches Institut
Hydroxylammoniumsulfat, Ergebnis der gewerbetoxikologischen Vorprüfung
Unpublished report (1969)

BASF AG, Gewerbehygiene und Toxikologie
Hydroxylammoniumchlorid, Ergebnis der gewerbetoxikologischen Vorprüfung
Unpublished report (1975)

BASF AG, Gewerbehygiene und Toxikologie
Hydroxylammoniumsulfat, Prüfung der dermalen Toxizität an der Rückenhaut weißer Kaninchen
Unpublished report (1980)

BASF AG, Gewerbehygiene und Toxikologie
Hydroxylammoniumsulfat, Prüfung der methämoglobinbildenden Wirkung an Katzen
Unpublished report (1981)

BASF AG, Abteilung Toxikologie
Bericht über die Prüfung der Toxizität von Hydroxylaminsulfat an Ratten bei 4wöchiger Gabe über das Trinkwasser (Range-Finding-Versuch)
Project No. 22C0389/8429 (1989 a)

BASF AG, Abteilung Toxikologie
Bericht über die Prüfung der oralen Toxizität von Hydroxylaminsulfat an Ratten, Verabreichung im Trinkwasser über 3 Monate
Project No. 32C0389/8474 (1989 b)

Bautz-Freese, E., Freese, E.
Two separable effects of hydroxylamine on transforming DNA
Proc. Natl. Acad. Sci. USA, 52, 1289–1297 (1964)

Bendich, A., Borenfreund, E., Korngold, G., Krim, M.
Action of hydroxylamine on DNA and chromosomes
Fed. Proc. Am. Soc. Exp. Biol., 22, 582 (1963)

Bernheim, M.L.C.
The reduction of hydroxylamine and some aryl hydroxamates by liver mitochondria from mammals and birds
Enzymologia, 43, 167–176 (1972)

Bhattacharya, A.K., Biswas, A., Haldar, G.
Genotoxicity of amino-acids, a probe
Indian J. Exp. Biol., 24, 460–463 (1986)

Binz, C.
Toxikologisches über das Hydroxylamin
Virchows Arch., 113, 1–9 (1888)

Borenfreund, E., Krim, M., Bendich, A.
Chromosomal aberrations induced by hyponitrite and hydroxylamine derivatives
J. Natl. Cancer Inst., 32, 667–679 (1964)

Brøgger, A.
Chromosome damage in human cells induced by hydroxylamine
Hereditas, 69, 19–24 (1971)

Brusick, D.J., Stetka, D.
Mutagenic assessment of hydroxylamine sulfate in the mouse micronucleus assay
Proc. Symp. Ind. Approach Chem. Risk Assess., 229–232 (1984)

Budovskii, E.I.
The mechanism of the mutagenic action of hydroxylamines
Prog. Nucleic Acid Res. Mol. Biol., 16, 125 (1976)

Chaube, S., Murphy, M.L.
The effects of hydroxyurea and related compounds on the rat fetus
Cancer Res., 26, 1448–1457 (1966)

Commercial Solvents Corporation
Toxicology of CSC-Products, Office Memorandum (1960)

Deguchi, T.
Activation of guanylate cyclase in cerebral cortex of rat by hydroxylamine
J. Biol. Chem., 252, 596–601 (1977)

Derelanko, M.J., Gad, S.C., Gavigan, F.A., Babich, P.C., Rinehart, W.E.
Toxicity of hydroxylamine sulfate following dermal exposure: variability with exposure method and species
Fundam. Appl. Toxicol., 8, 583–594 (1987)

DeSesso, J.M.
Demonstration of the embryotoxic effects of hydroxylamine in the New Zealand white rabbit
Anat. Rec., 196, 45A–46A (1980)

DeSesso, J.M.
Embryonic cell death caused by hydroxyurea and three other hydroxylamine-containing compounds
Teratology, 35 (2), 43A (1987)

DeSesso, J.M., Goeringer, G.C.
Developmental toxicity of hydroxylamine: an example of maternaly mediated effect
Toxicol. Industr. Health, 6 (1), 109–121 (1990)

Dunkel, V.C., Pienta, R.J., Sivak, A., Traul, K.A.
Comparative neoplastic transformation responses of Balb/3T3 cells, Syrian hamster embryo cells, and Rauscher murine leukaemia virus-infected Fischer 344 rat embryo cells to chemical carcinogens
J. Natl. Cancer Inst., 67, 1303–1315 (1981)

Engel, W., Krone, W., Wolf, U.
Die Wirkung von Thioguanin, Hydroxylamin und 5-Bromdesoxyuridin auf menschliche Chromosomen in vitro
Mutat. Res., 4, 353–368 (1967)

Epstein, S.S., Röhrborn, G.
Recommended procedures for testing genetic hazards from chemicals, based on the induction of dominant lethal mutations in mammals
Nature, 230, 459–460 (1971)

Epstein, S.S., Arnold, E., Andria, J., Bass, W., Bishop, Y.
Detection of chemical mutagens by the dominant lethal assay in the mouse
Toxicol. Appl. Pharmacol., 23, 288–325 (1972)

Evarts, R.P., Brown, C.A.
Morphology of mammary gland, ovaries, and pituitary gland of hydroxylamine-fed C3H/HeN mice
Lab. Invest., 37, 53–63 (1977)

Evarts, R.P., Brown, C.A., Atta, G.J.
The effect of hydroxylamine on the morphology of the rat mammary gland and on the induction of mammary tumours by 7,12-dimethylbenz(a)anthracene
Exp. Mol. Pathol., 30, 337–348 (1979)

Fahmy, C.G., Fahmy, M.J.
Hydroxylamine and derivatives: cytotoxicity without mutagenicity in cellular genetic systems
Chem. Biol. Interact., 2, 331–348 (1970)

Folesky, H., Nickel, H., Rothe, A., Zschunke, E.
Allergic eczema caused by hydroxylamine salts (oxammonium)
Z. Gesamte Hyg., 17, 353–356 (1971)

Freese, E., Bautz, E., Bautz-Freese, E.
The chemical and mutagenic specificity of hydroxylamine
Proc. Natl. Acad. Sci. USA, 47, 845–855 (1961)

Freese, E.
Molecular mechanism of mutations
In: Molecular Genetics, H. Taylor (ed.), part 1, chapter 5, p. 207–269, Academic Press, New York and London (1963)

Freese, E., Bautz-Freese, E.
The oxygen effect on deoxyribonucleic acid inactivation by hydroxylamine
Biochemistry, 4, 2419–2433 (1965)

Gad, S.C., Dunn, B.J., Dobbs, D.W., Reilly, C., Walsh, R.W.
Development and validation of an alternative dermal sensitization test: the mouse ear swelling test (MEST)
Toxicol. Appl. Pharmacol., 84, 93–114 (1986)

Gobbi, A.
Topic eczema due to hydroxylamine in workers engaged in the production of cycloserine
Med. Lav., 61, 387–391 (1970)

Goh, C.L.
Allergic contact dermatitis and onycholysis from hydroxylamine sulfate in colour developer
Contact Dermatitis, 22, 109 (1990)

Graf, U., Juon, H., Katz, A.J., Frei, H.J., Würgler, F.E.
A pilot study on a new Drosophila spot test
Mutat. Res., 120, 233–239 (1983)

Grant, W.M.
Toxicology of the eye
2nd ed., p. 569
Thomas, C.C. (ed.), Springfield, Illinois (1974)

Gross, P., Smith, R.P.
Biological activity of hydroxylamine: a review
Crit. Rev. Toxicol., 14 (1), 87–99 (1985)

Gupta, P., Sharma, T.
Non-random distribution of aberrations and identification with C- and G-bandings of the position of breakage points of muntjac chromosomes induced by mitomycin C, bromodeoxyuridine and hydroxylamine
Mutat. Res., 81, 63–74 (1981)

Gupta, P., Sharma, T.
Effect of mitomycin C, hydroxylamine and bromodeoxyuridin on cell cycle kinetics and sister chromatid exchange frequency of muntjac lymphocytes
Indian J. Exp. Biol., 20, 726–730 (1982)

Harman, D.
Prolongation of the normal lifespan and inhibition of spontaneous cancer by antioxidants
J. Gerontol., 16, 247–254 (1961)

ICI (Imperial Chemical Industries)
FYI-OTS-0484-0290, Supplement Sequence 1, March 1984

ILO (International Labor Office), Geneva
L. Parmeggiani (ed.): Encyclopaedia of Occupational Health and Safety
Vol. 1, 3rd (revised) ed., pp. 1091/1092 (1983)

Jacobsen, E., Plum, C.M., Milwertz, I.
Den eskperimentelle kroniske Hydroxylaminforgiftning
Nord. Med., 3, 2851–2858 (1939)

Jain, H.K., Shukla, P.T.
Locus specificity of mutagens in Drosophila
Mutat. Res., 14, 440–442 (1972)

Kao, F.T., Puck, T.T.
Genetics of somatic mammalian cells
IX. Quantitation of mutagenesis by physical and chemical agents
J. Cell. Physiol., 74, 245–258 (1969)

Kimball, R.F., Hirsch, B.F.
Tests for the mutagenic action of a number of chemicals on *Haemophilus influenzae* with special emphasis on hydrazine
Mutat. Res., 30, 9–20 (1975)

Klausmeier, R.E., Bard, R.C.
Ammonium dehydrogenase
J. Bacteriol., 68, 129–130 (1954)

Kruszyna, R., Kruszyna, H., Smith, R.P.
Comparison of hydroxylamine, 4-dimethylaminophenol and nitrite protection against cyanide poisoning in mice
Arch. Toxicol., 49, 191–202 (1982)

Kruszyna, R., Kruszyna, H., Doherty, P., Smith, R.P.
Hypotensive effects of hydroxylamine in intact anaesthetized dogs and cats
Arch. Toxicol., 55, 203–205 (1984)

Lewin, L.
XVII. Über Hydroxylamin. Ein Beitrag zur Kenntnis der Blutgifte
Arch. Exp. Pathol. Pharmacol., 25, 306–325 (1889)

Lewis, R.J., sr., Tatken, R.L. (eds.)
Registry of Toxic Effects of Chemical Substances (1980) ed. U.S. Department of Health and Human Services (1982) 913
Cited in: Gross, P., Smith, R.P. (1985)

Litton Bionetics Inc.
Mutagenicity evaluation of hydroxylamine sulfate in the mouse micronucleus test
Final Report, LBI Project No. 21064, May 1979, Revised April 1980

Mahna, S.K.
Transfection assay test system: assay of four chemical mutagens
Indian J. Exper. Biol., 22, 338–339 (1984)

Makotcenko, V.M., Rutstejn, L.P.
Eine Vergiftung mit Hydroxylaminhydrochlorid (deutsche Übersetzung TIB/U-7270)
Vrach. Delo., 7, 125–127 (1974)

Malle, K.G.
Hydroxylamin
In: Ullmanns Encyklopädie der technischen Chemie
4th ed., vol. 13, p. 169–172
Verlag Chemie, Weinheim (1977)

Malling, H.V., de Serres, F.J.
Hydroxylamine-induced purple adenine (ad-3) mutants in *Neurospora crassa*
I. Characterization of mutants by genetic tests
Mutat. Res., 12, 35–46 (1971)

McCann, J., Choi, E., Yamasaki, E., Ames, B.N.
Detection of carcinogens as mutagens in the Salmonella/microsome test assay of 300 chemicals
Proc. Natl. Acad. Sci. USA, 72, 5135–5139 (1975)

Menevse, S., Kustimur, S., Menevse, A., Ekmekci, A., Ergin, M.
Salmonella tester strain TA 104, for the detection of mutagenic and carcinogenic chemicals in our environment
Mikrobiyol Bült, 18, 99–106 (1984)

Microbiological Associates, Bethesda, Maryland
Activity of T1680 in the cell transformation assay in the absence of exogenous metabolic activation
Final Report for BASF AG (1981 a)

Microbiological Associates, Bethesda, Maryland
Activity of T1680 in the cell transformation assay in the presence of exogenous metabolic activation
Final Report for BASF AG (1981 b)

Mironescu, S.T., Burducea, O., Sahnazarov, N.
Nucleolar and mitotic abnormalities produced by thioacetamide and hydroxylamine in monkey kidney cells cultivated in vitro
Exp. Cell Res., 57, 193–204 (1969)

Murphy, M.L., Chaube, S.
Preliminary survey of hydroxyurea (NSC-32065) as a teratogen
Cancer Chemother. Rep., 40, 1–7 (1964)

Myhr, B.C., Caspary, W.J.
Evaluation of the L5178Y mouse lymphoma cell mutagenesis assay: interlaboratory results for sixty three coded chemicals tested at Litton Bionetics Inc.
Environ. Mol. Mutagen., 12 (Suppl. 13), 103–194 (1988)

NIH/NTP, National Institute of Health, National Toxicology Program
FYI-OTS-0284-0290, Suppl. Seq. B. Letter re. National Cancer Institute/NTP testing of hydroxylamine hydrochloride from D.A. Canter (NTP) to T. O'Bryan
Office of Toxic Substances, U.S. Environmental Protection Agency (1984)

Oil and Hazardous Materials Assistance Data System (1978)
Cited in: Gross, P. and Smith, R.P. (1985)

Oppenheim, J.J., Fishbein, W.N.
Induction of chromosome breaks in cultured normal human leucocytes by potassium arsenite, hydroxyurea and related compounds
Cancer Res., 25, 980–985 (1965)

OTS (Office of Toxic Substances)
Chemical Hazard Information Profile (CHIP) Hydroxylamine, Hydroxylamine hydrochloride, Hydroxylamine sulfate (2:1), Hydroxylamine sulfate (1:1) (1984)

Parkash, R., Miglani, G.S.
Hydroxylammoniumsulfate and maleic hydrazide induced chromosomal aberrations in Drosophila
Caryologia, 31, 1–5 (1978)

Pellerat, M., Chabeau, G.
Hydroxylamine et dermatoses professionelles
Société Francaise de Dermatologie et de Syphiligraphie, Saint-Etienne meeting, Session of 11 October 1975

Pienta, R.J.
Transformation of Syrian-hamster embryo cells by diverse chemicals and correlation with their reported carcinogenic and mutagenic activities
Chem. Mutagens, 6, 175–202 (1980)

Putrament, A., Baranowska, H.
Induction of intragenic mitotic recombination in yeast by hydroxylamine and its mutagenic specificity
Mol. Gen. Genet., 111, 89–96 (1971)

Raimondi, C., Bertoni, G.
Sull'azione tossica dell'idrosilamina
Annali Universali di Med., 259, 97 (1882)
Cited in: Binz (1888)

Riemann, H.
On the toxicity of hydroxylamine
Acta Pharmacol., 6, 285–292 (1950)

Rosenkranz, H.S., Poirier, L.A.
Evaluation of the mutagenicity and DNA-modifying activity of carcinogens and non carcinogens in microbial systems
J. Natl. Cancer Inst., 62 (4), 873–891 (1979)

Runnegar, M.T.C., Blakeley, R.L., Fernley, R.T., Webb, E.C., Zerner, B.
The system: hydroxylamine-oxygen-cupric ion. A model hydroxylase in the inactivation of ox and pig liver carboxylesterases
Biochem. Biophys. Acta, 167, 632–634 (1968)

Saul, R.L., Archer, M.C.
Oxidation of ammonia and hydroxylamine to nitrate in the rat
In: O'Neill, I.K. et al. (ed.)
IARC Scientific Publications, No. 57
N-Nitroso compounds: occurrence, biological effects and relevance to human cancer; 8th International Symposium,
Banff., Canada, Sept. 5–9, 1983, XX + 1011 P., Oxford University Press: New York, N.Y., USA; pp. 241–246 (1984)

Savolainen, H.
Neurochemical effects of ingested hydroxylamine
Res. Commun. Chem. Pathol. Pharmacol. 46, (2), 275–281 (1984)

Shukla, P.T.
Analysis of mutagen specificity in *Drosophila melanogaster*
Mutat. Res., 16, 363–371 (1972)

Shukla, P.T., Auerbach, C.
The delayed mutagenic action of hydroxylamine in Drosophila
Mutat. Res., 61, 399–400 (1979)

Siegel, L.M., Leinweber, F.J., Monty, K.J.
Characterization of the sulfite and hydroxylamine reductases of Neurospora crassa
J. Biol. Chem., 240, 2705–2711 (1965)

Simmon, V.F.
In vitro assays for recombinogenic activity of chemical carcinogens and related compounds with *Saccharomyces cerevisiae* D3
J. Natl. Cancer Inst., 62 (4), 901–909 (1979 a)

Simmon, V.F.
In vitro mutagenicity assays of chemical carcinogens and related compounds with *Salmonella typhimurium*
J. Natl. Cancer Inst., 62, 893–899 (1979 b)

Smith, R.P., Layne, W.R.
A comparison of the lethal effects of nitrite and hydroxylamine in the mouse
J. Pharmacol. Exp. Ther., 165, 30–35 (1969)

Sokova, O.I., Kulagina, O.E., Pogosyants, E.E.
Study of the action of gamma rays, hydroxylamine and urethane on the chromosomes of the Dzungarian hamster in vitro
Sov. Genet., 6, 1593–1597 (1970)

Somers, C.E., Hsu, T.C.
Chromosome damage induced by hydroxylamine in mammalian cells
Proc. Natl. Acad. Sci. USA, 48, 937–943 (1962)

Speit, G., Wick, C., Wolf, M.
Induction of sister chromatid exchanges by hydroxylamine, hydrazine and isoniazid and their inhibition by cysteine
Hum. Genet., 54, 155–158 (1980)

Stenbaeck, J., Weisburger, J.H., Williams, G.M.
Hydroxylamine effects on cryptogenic neoplasm development in C3H mice
Cancer Lett., 38, 73–85 (1987)

Stoll, R., Bodit, F., Maraud, R.
Sur l'action tératogène de l'hydrazine et des substance voisines chez l'embryon de poulet
C.R. Soc. Biol. (Bordeaux), 161, 1680–1684 (1967 a)

Stoll, R., Bodit, F., Maraud, R.
Sur l'action de divers antimitotiques apparentés à l'hydroxyurée et à la semicarbazide chez l'embryon de poulet
C.R. Soc. Biol. (Bordeaux), 161, 1963–1967 (1967 b)

Taylor, W.D., Mudgett, J.S., Bonneau, R., Wilt, S., Kroekel, N., Brontz, P.C., Heicklen, J.
Toxic and mutagenic effects on bacteria and mammalian cells of hydroxylamines
Environ. Mutagen., 7, Suppl. 3, 86 (1985)

Tessman, I., Ishiwa, H., Kumar, S.
Mutagenic effects of hydroxylamine in vivo
Science, 148, 507–508 (1965)

Traul, K.A., Takayama, K., Kachevsky, V., Hink, R.J., Wolff, J.S.
A rapid in vitro assay for carcinogenicity of chemical substances in mammalian cells utilizing an attachment-independence endpoint
J. Appl. Toxicol., 1, 190–195 (1981)

Williams, G.M., Laspia, M.F., Dunkel, V.C.
Reliability of the hepatocyte primary culture/DNA repair test in testing of coded carcinogens and non-carcinogens
Mutat. Res., 97, 359–370 (1982)

Winter, A.J., Hokanson, J.F.
Effects of long-term feeding of nitrate, nitrite, or hydroxylamine on pregnant dairy heifers
Am. J. Vet. Res., 25, 353–361 (1964)

Würgler, F.E., Graf, U., Frei, H., Juon, H., Katz, A.J.
Mutagenicity testing with somatic cells of *Drosophila melanogaster*
Mutat. Res., 113, 350–351 (1983)

Yamamoto, R.S., Weisburger, E.K., Korzis, J.
Chronic administration of hydroxylamine and derivatives in mice
Proc. Soc. Exp. Biol. Med., 124, 1217–1220 (1967)

Yoshikawa, S., Orii, Y.
The inhibition mechanism of the cytochrome oxidase reaction
I. Inhibition by hydroxylamine
J. Biochem., 68, 145–156 (1970)

Zachariah, P.K., Lee, Q.P., Symms, K.G., Juchau, M.R.
Further studies on the properties of human placental microsomal cytochrome P-450
Biochem. Pharmacol., 25, 793–800 (1976)

Zimmermann, W., Gottschewski, G.H.M.
Die Wirkung bestimmter Substanzen auf die DNS- und Proteinsynthese in der frühen Embryonalentwicklung des Kaninchens
Bull. Schweiz. Akad. Med. Wiss., 22, 166–183 (1966)

Zinc ethylphenyldithiocarbamate

1. Summary and assessment

Zinc ethylphenyldithiocarbamate reacts readily with nitrosating agents to give ethylphenylnitrosamine.

After oral administration of zinc ethylphenyldithiocarbamate to pregnant rats, aniline and N-ethyl aniline are detectable as metabolites in the blood. The maximum concentrations are reached after 24 hours at 10 and 2.01 μg/ml, respectively, and these decline after 3 days to less than 0.3 μg/ml. After the administration of a single oral dose to rats, carbon disulphide is detectable dose-dependently in the exhaled air.

Zinc ethylphenyldithiocarbamate is non-toxic after a single oral dose (LD_{50} rat >10 000 mg/kg body weight). In studies in pregnant rats, methaemoglobin formation occurs after oral administration, apparently due to the metabolites.

In rabbits, zinc ethylphenyldithiocarbamate is not irritating to either intact or scarified skin and is only slightly irritating to the eyes.

After subchronic oral administration to rats, 500 mg/kg for 1.5 months cause deaths and 100 and 10 mg/kg for 4 months lead to changes in the parameters studied (e.g. blood composition, prothrombin time, cytochrome oxidase activity, histology). No such effects are evident after 10 months at 1 mg/kg.

Zinc ethylphenyldithiocarbamate induces point mutations in the Salmonella/microsome test in strains TA 1535 and TA 100 without metabolic activation at concentrations of 100 μg/plate and above and with metabolic activation at 500 μg/plate and above.

Zinc ethylphenyldithiocarbamate is embryotoxic and teratogenic to hens' embryos by an administration route that is not relevant in practice, in a protocol that does not correspond to present-day requirements. In rats, zinc ethylphenyldithiocarbamate causes dose-dependent foetal deaths only at maternally toxic doses of 62.5 mg/kg or more. At this dose range it also increases skeletal variations, but causes no skeletal or soft tissue malformations. In a subsequent study with higher doses of up to 1000 mg/kg, it is confirmed that the foetotoxic activity of zinc ethylphenyldithiocarbamate in rats is the direct result of maternal toxicity.

In man, zinc ethylphenyldithiocarbamate can, like disulfiram, cause hypersensitive reactions to alcohol, but reference to this is perhaps only contained in safety data sheets as a precaution. Skin sensitization due to zinc ethylphenyldithiocarbamate in humans appears to be relatively rare according to experience in the rubber industry to date.

2. Name of substance

2.1	Usual name	Zinc ethylphenyl-dithiocarbamate
2.2	IUPAC-name	Bis (N-ethyldithio-carbanilato)-zinc
2.3	CAS-No.	14634-93-6

3. Synonyms, common and trade names

Bis (ethylphenyl-carbamodithio-ato-S,S')-zinc
Vulkacit P extra N
Zinkethylphenyldithiocarbamat

4. Structural and molecular formulae

4.1 Structural formula

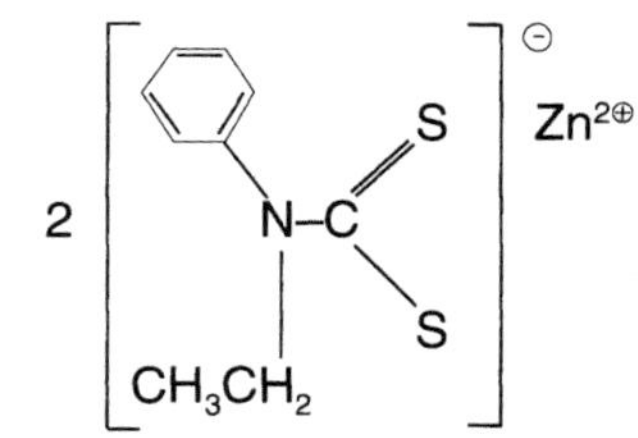

4.2 Molecular formula $C_{18}H_{20}N_2S_4Zn$

5. Physical and chemical properties

5.1	Molecular mass, g/mol	457.99
5.2	Melting point, °C	⩾200 (WTR, 1984; Bayer, 1988)
5.3	Boiling point, °C	–
5.4	Vapour pressure, hPa	No information available
5.5	Density, g/cm³	Ca. 1.5 (Bayer, 1988)
5.6	Solubility in water	Insoluble (WTR, 1984; Bayer, 1988)
5.7	Solubility in organic solvents	Soluble in chloroform and toluene Readily soluble in methylene chloride Dissolves with difficulty in acetone and ethyl acetate Insoluble in petrol and ethanol (WTR, 1984; Bayer, 1986, 1988)
5.8	Solubility in fat	No information available
5.9	pH-value	–
5.10	Conversion factor	1 ppm ≙ 18.69 mg/m³ 1 mg/m³ ≙ 0.05 ppm (at 25 °C and 1013 hPa)

6. Uses

Vulcanization accelerator in the rubber industry (Bayer, 1986).

7. Experimental results

Like other amine-based dithiocarbamates, zinc ethylphenyldithiocarbamate reacts readily with nitrosating agents to form the corresponding nitrosamine, ethylphenyl nitrosamine, which has been shown to be carcinogenic in animal studies. Technical grade products are, like other dithiocarbamates, contaminated with the nitrosamine (Spiegelhalder and Preussmann, 1983).

7.1 Toxicokinetics and metabolism

Metabolism studies were carried out to investigate the methaemoglobinaemia observed in pregnant rats after exposure to zinc ethylphenyldithiocarbamate. Following oral administration of 1000 mg/kg body weight to pregnant Wistar rats, N-ethylaniline and aniline were identified (GC-MS) as metabolites in the blood by extraction with ethyl acetate. Maximun concentrations of N-ethylaniline (10 µg/ml) and aniline (2.01 µg/ml) were found in the blood 24 hours after administration. These concentrations dropped to less than 0.3 µg/ml within 3 days. N-Ethylaniline and aniline were also detected after oral administration of N-ethylaniline (125 mg/kg) to rats, with maximum concentrations (5.52 and 1.25 µg/ml, respectively) reached after 8 hours. These declined to less than 0.2 µg/ml within 12 hours. The increase in blood methaemoglobin after administration of zinc ethylphenyldithiocarbamate or N-ethylaniline was almost proportional to the decrease in the amine concentration in the blood. A comparison of the effects of zinc ethylphenyldithiocarbamate, N-ethylaniline and aniline on blood composition in rats (erythrocytes, methaemoglobin, haemoglobin, haematocrit value) revealed that these substances behaved similarly with regard to the induction of methaemoglobinaemia. It was concluded from these results that the methaemoglobinaemia observed after exposure to zinc ethylphenyldithiocarbamate was produced by the metabolites N-ethylaniline and aniline (Nakaura et al., 1985).

Rats were given single oral doses of 1500, 100 or 50 mg zinc ethylphenyldithiocarbamate (Vulkacit)/kg body weight. The carbon disulphide content in the exhaled air was then determined for 9 days. After 1500 mg/kg, a total of 224 µg carbon disulphide was excreted within 8 days, after 100 mg/kg a total of 43 µg was excreted within 2 days, while after 50 mg/kg, no carbon disulphide was detected (Yalkut, 1971).

7.2 Acute and subacute toxicity

10 Male and 10 female Wistar rats (174 to 250 g and 97 to 165 g, respectively) were given a single dose of 25 ml of a 40%

(W/V) suspension of zinc ethylphenyldithiocarbamate in propylene glycol. The LD_{50} value was more than 10 000 mg/kg body weight. Signs of toxicity seen in these animals a few hours after treatment included pale hind legs and dark eyes. During the first week after treatment, the edges of the eyes and the nostrils were encrusted. 1 Female and 4 males died between the 2nd and 5th day after administration. In the 2nd week of the 14-day observation period the surviving rats gradually recovered, but they were still not in optimum condition at the end of the study. Dark staining of the spleen and kidneys was observed at autopsy in all of the rats that survived the treatment (van Beek et al., 1975).

7.3 Skin and mucous membrane effects

In order to study the primary skin irritancy of zinc ethylphenyldithiocarbamate, 500 mg were applied to the intact skin of 6 adult New-Zealand rabbits and to the scarified skin of 6 further rabbits, under a patch (fixed with adhesive tape) for 24 hours. The results were evaluated 24 and 72 hours after application, according to the Draize method (1944), but skin irritation was not seen in any of the treated rabbits (van Beek et al., 1975).

Eye irritation tests with zinc ethylphenyldithiocarbamate were likewise carried out on 6 New Zealand rabbits. In this study, 100 mg of the substance were instilled into the conjunctival sac of one eye. The other untreated eye served as a control. Evaluation of the findings was carried out 24, 48 and 72 hours and also 7 days after instillation, according to the methods of the FDA (1965) and Draize and Kelley (1952). 24 Hours after instillation, slight damage to the cornea was observed in 1 of the 6 rabbits, while slight to moderate conjunctivitis was observed in all rabbits. The effects gradually began to clear up, but were not completely reversed, so that one day 7, slight redness of the conjunctiva could still be seen. Based on these results, zinc ethylphenyldithiocarbamate was considered as an eye irritant (van Beek et al., 1975).

7.4 Sensitization

No information available.

7.5 Subchronic and chronic toxicity

In a study of the subchronic toxicity of zinc ethylphenyldithiocarbamate (Vulkacit P extra N), groups of 10 white rats (strain and sex not specified) were treated 6 times a week by stomach tube at dose levels of 500 mg/kg for 1.5 months, 100 mg/kg or 10 mg/kg for 4 months, or 1 mg/kg for 10 months. After treatment with 500 mg/kg, 3 rats died after 32, 34 and 41 days. The other doses did not cause any deaths. Body weight gain after treatment with 500 mg/kg for 1.5 months was 12% lower than the control value, while after treatment with 100 mg/kg for 2 months it was 16% lower than the control. Both high doses caused marked disturbances in blood composition, and at the 10 mg/kg dose there was a trend towards a reduction in the erythrocyte count at the end of the third month of treatment, and marked reticulocytosis at the end of the study. Furthermore, the prothrombin time was reduced after 3 months at 100 mg/kg and 4 months at 10 mg/kg. Liver cytochrome oxidase activity was reduced ($p<0.01$) after the administration of 100 mg/kg for 4 months, while ceruloplasmin activity was increased ($p<0.01$) after the administration of 100 mg/kg for 4 months or 10 mg/kg for 3 months. The catalase-index was increased ($p<0.05$) after administration of 100 mg/kg for 2 and 4 months. A test of thyroid gland function with ^{131}I did not reveal any deviations from control values. Histologically, there was slight granular degeneration of the liver cells at the 100 and 10 mg/kg dose levels. Other effects included an increase in cells of an alveolar or histiocytic nature in the lungs, peribronchial infiltration, dilation and hyperaemia of the kidney vessels, granular degeneration of the epithelium of the tubules with isolated necroses, and dilated and blocked vessels in the spleen. At the 100 mg/kg dose level there were also numerous megakaryocytes in the spleen pulp. Oral administration of 1 mg/kg for 10 months had no overt effects. The author concluded from these results that zinc ethylphenyldithiocarbamate is of low toxicity compared with the majority of thiocarbamates (Yalkut, 1971).

No toxic effects were observed after mice were given rubber extract to drink (duration not specified) which contained 2.5 mg Zn^{2+}, 0.14 mg Vulkacit P extra N (zinc ethylphenyldithiocarbamate) and 0.02 mg thiuram EF/L (Khoroshilova et al., 1980).

7.6 Genotoxicity

7.6.1 In vitro

The ability of zinc ethylphenyldithiocarbamate to induce point mutations was studied in the Salmonella/microsome test at concentrations of 50, 100, 250 and 500 μg/plate in strains TA 1535, TA

1537, TA 1538, TA 98 and TA 100. A positive effect was found in strains TA 1535 and TA 100 from 500 µg/plate with metabolic activation (S9-fraction from the livers of Arochlor 1254 pre-treated male Sprague-Dawley rats), while without a metabolizing system a positive result was obtained at 100 µg/plate. Thus zinc ethylphenyldithiocarbamate proved to be a direct-acting mutagen (Hedenstedt et al., 1979; Hedenstedt, 1981).

In another study with the *Salmonella typhimurium* strains TA 1535, TA 1537, TA 1538, TA 98 and TA 100, the lowest effective dose was likewise 100 µg/plate with and without metabolic activation (Rannug et al., 1984).

7.6.2 In vivo

No information available.

7.7 Carcinogenicity

The oral administration of a rubber extract containing 2.5 mg Zn^{2+}, 0.14 mg Vulkacit P extra N (zinc ethylphenyldithiocarbamate) and 0.02 mg thiuram EF/L to mice in their drinking water did not lead to the induction of cancer (Khoroshilova et al., 1980). No further details were available in the English abstract of this work, or in the summary in Chemical Abstracts.

7.8 Reproductive toxicity

The effects of zinc ethylphenyldithiocarbamate (technical grade) on the development of 3-day-old chicken embryos (White Leghorn) have been investigated. Doses of 2.5, 5.0, 7.5, 10.0 and 20.0 nM/egg in acetone (administered in volumes of 5 µl) were dropped into the air chamber onto the embryo (ca. 30 eggs per concentration) through a window that had been cut in the egg after 3 days (72 to 76 hours) of incubation. After treatment, the window was sealed with transparent adhesive tape. 10 Eggs treated with 5 µl acetone served as controls. The eggs were candled every 2 to 3 days. Eggs that contained dead embryos were opened to determine the stage of development (as defined by Hamburger and Hamilton, 1951) as well as possible external malformations. Incubation was terminated on day 14 (11 days after administration). The remaining eggs were opened and the embryos checked for survival and external malformations. The embryotoxic potency was estimated from the total number of embryos affected 11 days after administration, taking into consideration total mortality and the number of malformed embryos. In total, of the treated embryos, 13% were affected at 2.5 nM, 65% at 5 nM, 88% at 7.5 nM and 100% at 10.0 and 20.0 nM (excluding mortality in the early development phase). Over the period of 11 days after administration, an ED_{50} of 4.3 nM/egg was calculated. The LD_{50} war reported as 5.2 nM/egg. The highest proportion of embryos with malformations (63%) was found after treatment with 7.5 nM/egg. These malformations included two types of eye defect (small eye cups and partial or total absence of lids and cornea), anomalies of the wings and shoulders, defects of the coelomic wall and anomalies of the neck and spinal column. Based on these results the authors evaluated zinc ethylphenyldithiocarbamate as a potent embryotoxic and teratogenic compound (Korhonen, 1981; Korhonen et al., 1983 a, b).

To study the effects of zinc ethylphenyldithiocarbamate on the pre- and postnatal development of mammals, doses of 31.2, 62.5, 125 and 250 mg/kg suspended in olive oil were given to pregnant rats (numbers not specified) by stomach tube from day 7 to day 15 or pregnancy. The authors reported that increasing doses caused an increase in foetal mortality, and 250 mg/kg was lethal to all feotuses. Although increased early death, reduced body weight and skeletal variations were observed in the foetuses of dams treated with 62.5 and 125 mg/kg, there were no indications of skeletal or visceral malformations which could be attributed to treatment with zinc ethylphenyldithiocarbamate in any group. After treatment with doses of 62.5 mg/kg and above, the body weights were significantly reduced in the pregnant rats, while significantly increased spleen weights were seen from 31.2 mg/kg. The foetotoxic effects were therefore considered to be connected with the maternal toxicity (Nakaura et al., 1983).

In a further experiment in which rats were given 125, 250, 500 or 1000 mg/kg daily from day 7 to day 9 of pregnancy, the same authors reported a reduction in erythrocytes, haemoglobin content and haematocrit value in the pregnant rats, as well as an increase in microcytes. From the results of both studies, the authors concluded that under the experimental conditions specified, zinc ethylphenyldithiocarbamate induced a toxic, but not teratogenic, effect and that the early death was attributable to the effects of zinc ethylphenyldithiocarbamate on the erythrocytes of the pregnant rats (Nakaura et al., 1983).

7.9 Effects on the immune system

No information available.

7.10 Neurotoxicity

No information available.

7.11 Other effects

No information available.

8. Experience in humans

Two safety data sheets indicate that, like disulfiram, zinc ethylphenyldithiocarbamate can cause hypersensitivity to alcohol (WTR, 1984; Bayer, 1988).

Zinc ethylphenyldithiocarbamate was studied at a concentration of 1% in a patch test in order to identify allergens in rubber products. Evaluation of the results 72 hours after application showed a positive reaction in 2 out of 31 patients (6%) with rubber-induced contact dermatitis (no further information was contained in the English summary of this Japanese work; Kantoh et al., 1985).

Zinc ethylphenyldithiocarbamate was tested in 2569 eczema sufferers in the course of a study on the sensitizing effect of vulcanization accelerators. Hypersensitivity reactions to this substance were observed in less than 0.5% of the patients studied. In the English summary of this Polish work, however, there were no details of the basic study conditions (no exact description of the concentrations studied, the test conditions or test methods; Rudzki and Napiorkowska, 1984).

Ethylphenyl nitrosamine was detectable in the air from production units when zinc ethylphenyldithiocarbamate was used as a vulcanization accelerator in the rubber industry (Spiegelhalder and Preussmann, 1983).

References

Bayer AG, Kautschuk-Chemikalien
1. Vulkanisationschemikalien
Technisches Informationsblatt (1986)

Bayer AG, Geschäftsbereich Kautschuk
DIN-Sicherheitsdatenblatt Vulkacit P extra N (1988)

Draize
J. Pharmcol., 82, 377–390 (1944)
Cited in: van Beek et al. (1975)

Draize, Kelley
Drug Cosmet. Ind., 71, 36 (1952)
Cited in: van Beek et al. (1975)

FDA (US Food and Drug Administration)
Fed. Reg., 28 (119), 5582 (1963)
Cited in: van Beek et al. (1975)

Hamburger, V., Hamilton, H.L.A.
A series of normal stages in the development of the chick embryo
J. Morphol., 88, 49–92 (1951)
Cited in: Korhonen et al. (1983 a)

Hedenstedt, A., Rannug, U., Ramel, C., Wachtmeister, C.A.
Mutagenicity and metabolism studies of 12 thiuram and dithiocarbamate compouds used as accelerators in the Swedish rubber industry
Mutat. Res., 68, 313–325 (1979)

Hedenstedt, A.
Mutagenicity of rubber-vulcanization gases
SGF (Sver. Gummitek. Foeren.) Publ., 57, 39–46 (1981)

Kantoh, H., Ishihara, M., Itoh, M., Hosono, K., Nishimura, M.
Allergens in rubber products
HIFU (Skin Research), 27, 501–509 (1985)

Khoroshilova, N.V., Sal'nikova, L.S., Efremova, V.M.
Biological and sanitary-chemical properties of a heat-resistant rubber made from SKN-26M with the vulcanization accelerator thiuram EF

Kauch. Rezina., 10, 54–55 (1980)
Chem. Abstr., No. 916 T, 94 (1981)

Korhonen, A.
Toxicity and teratogenicity of thiurams and dithiocarbamates in the chick embryo test
Proceedings of the Second Finnish-Estonian Symposium on Early Effects of Toxic Substances, 64–71 (1981)

Korhonen, A., Hemminki, K., Vainio, H.
Embryotoxicity of industrial chemicals on the chicken embryo: dithiocarbamates
Teratogenesis Carcinog. Mutagen., 3, 163–175 (1983 a)

Korhonen, A., Hemminki, K., Vainio, H.
Toxicity of rubber chemicals towards three-day chicken embryos
Scand. J. Work Environ. Health, 9, 115–119 (1983 b)

Nakaura, S., Tanaka, S., Kawashima, K., Takanaka, A., Omori, Y.
Teratological studies of zinc ethylphenyldithiocarbamate (ZEPC), a vulcanization accelerator for rubber, in rats
J. Toxicol. Sci., 8, 337 (1983)

Nakaura, S., Kawanishi, T., Ohno, Y., Kawashima, K., Tanaka, S., Takahashi, S., Takanaka, A., Omori, Y., Matumoto, K.
Fetal toxicity of zinc ethylphenyldithiocarbamate (ZEPC) (3). Mechanism of formation of methaemoglobin caused by ZEPC
J. Toxicol. Sci., 10, 252 (1985)

Rannug, A., Rannug, U., Ramel, C.
Genotoxic effects of additives in synthetic elastomers with special consideration to the mechanism of action of thiurams and dithiocarbamates
Prog. Clin. Biol. Res., 141, ISS. Ind. Hazards Plast. Synth. Elastomers, 407–419 (1984)

Rudzki, E., Napiorkowska, T.
Sensitization by carbamate accelerators of rubber cure
Przeg. Derm., 71, 121–125 (1984)

Spiegelhalder, B., Preussmann, R.
Occupational nitrosamine exposure. 1. Rubber and tyre industry
Carcinogenesis, 4, 1147–1152 (1983)

Beek, van, L., Til, H.P., Spanjers, M.T.
Toxicological examination of the compound WTR 44
CIVO-TNO Report no. R 4848 (1975)

WTR (International Working Group on the Toxicology of Rubber Additives)
Safety data sheet zinc ethylphenyldithiocarbamate (1984)

Yalkut, S.I.
The toxicology of resin Vulcacite P extra N accelerator of vulcanization (English translation)
Gig. Sanit., 36, 28–31 (1971)

Compound chemical index vol. 1–6

Springer-Verlag and the Environment

We at Springer-Verlag firmly believe that an international science publisher has a special obligation to the environment, and our corporate policies consistently reflect this conviction.

We also expect our business partners – paper mills, printers, packaging manufacturers, etc. – to commit themselves to using environmentally friendly materials and production processes.

The paper in this book is made from low- or no-chlorine pulp and is acid free, in conformance with international standards for paper permanency.

MIX
Papier aus verantwortungsvollen Quellen
Paper from responsible sources
FSC® C105338

If you have any concerns about our products,
you can contact us on
ProductSafety@springernature.com

In case Publisher is established outside the EU,
the EU authorized representative is:
Springer Nature Customer Service Center GmbH
Europaplatz 3, 69115 Heidelberg, Germany

Printed by Libri Plureos GmbH
in Hamburg, Germany